AF342027

EXOPLANETS

EXTANT LIFE?

SPACE SCIENCE, EXPLORATION AND POLICIES

Additional books in this series can be found on Nova's website
under the Series tab.

Additional e-books in this series can be found on Nova's website
under the e-book tab.

SPACE SCIENCE, EXPLORATION AND POLICIES

EXOPLANETS

EXTANT LIFE?

DIETER REHDER

nova publishers
New York

NOTICE TO THE READER

Library of Congress Cataloging-in-Publication Data

Exoplanets : extant life?/Dieter Rehder (Chemistry Department, University of Hamburg, Germany), editor.
     pages cm. -- (Space science, exploration and policies)
  Includes bibliographical references and index.
  ISBN 978-1-63463-301-7 (hardcover)
  1. Life on other planets. 2. Extrasolar planets. 3. Habitable planets. 4. Exobiology. I. Rehder, Dieter, editor.
  QB54.E93 2014
  576.8'39--dc23
                        2014039719

Published by Nova Science Publishers, Inc. † New York

CONTENTS

LIST OF SIDEBARS

PREFACE

Barely half a millennium ago, the philosopher and astronomer Giordano Bruno (1548-1600) stated that "The Universe is infinite in time and space, containing, in addition to Earth, an infinite number of inhabited planets". Giordano became isolated from the Church in 1593, and was executed in 1600. 400 years later, Pope John Paul II included him in a general apology to philosophers and scientists condemned by the inquisition. The search for extraterrestrial life has in fact been a long-standing tradition, and discussions with respect to life beyond Earth have been revived ever since the discovery of the first exoplanets orbiting a Sun-like star (51 Pegasi-b) in 1995, and the evidence of ancient as well as for subsoil water on Mars provided in the past decade by the various Mars missions.

Perhaps the best-known activity in the search for extraterrestrial life is the SETI project. SETI stands for Search for Extraterrestrial Intelligence, and aims to detect alien civilizations via specific features in electromagnetic radiation, such as connected to radio broadcast. Even in case there is extraterrestrial intelligence somewhere in other planetary systems, and provided that their way of intercommunication compares to what we are accustomed to and exchange of facts is thus principally feasible, there are several caveats to the assumption that such a communication will ever come about: The chances that this extraterrestrial intelligence, and its achievements in communication, is on about the same level as ours is tiny to the extent that we can hardly expect to detect any such signaling. In the most optimistic scenario, our extraterrestrial counterparts are either a few hundred centuries behind our technical achievements, and thus incapable of radio communication, or they are a few hundred centuries in advance, and have adapted to communication on an alternative technical level (yet) unknown to us – or decided that there is no need at all to communicate with other potential civilizations. Furthermore, since the propagation velocity of radio signals is restricted to the speed of light, any information coming from systems hundreds or thousands of light years away from Earth will contain outdated information.

The reader will have noticed that "SETI" and related phantasms are not my cup of tea. I am certainly quite enthusiastic when it comes to Science Fiction, and my private library abounds from Dominique, Aldis, Asimov, van Voigt, Bradbury and the like. However, Science Fiction is one thing, and the scientifically-based elucidation of our cosmos – including the viability of life somewhere else – is another thing. More viable approaches to the existence and propagation of life within our Solar System and beyond have been disseminated, such as panspermia, or the Gaia hypothesis. Panspermia is a theory according to which germs of life have spread across space in between the planets (and planets + moons) of a solar system, and also in between different solar systems. Lovelock's Gaia hypothesis focuses on the planet Earth as a vital entity. These scenarios will briefly be appreciated in section 7.4 and in the Epilogue; they are, however, not the focus of this book.

Rather, the main intention of the present treatise is to delineate the formation of molecular species in interstellar clouds, and to elucidate those scenarios that likely have led, based on the chemical inventory of interstellar clouds, to the synthesis of complex molecules and molecular assemblies that are present for example on our home planet and presumably also on "Earth-like" exoplanets in a multitude of other solar systems. The book further aims towards the elucidation of the conditions by which such complex molecular assemblies might have initiated the formation of life molecules and the coming into existence – including the evolution, sustenance and survival – of life itself. Together with the "chemistry issue", I will to some extent elaborate on the astronomical background, as far as the generation and the fate of stars and the formation of planetary systems is concerned, with our Solar System as the basic point of reference.

The present book addresses readers with a basic knowledge of facts related to astronomy and to chemistry. In order to give the reader an impression of the complexity of the molecular structures and reaction paths that contribute to the constitution of Life, I will provide a set of basic axioms bearing on simple and complex chemical species involved in cooperative networks typical for life processes. The less scholarly reader is encouraged to simply skip these interjections, and to profit for the main part from the main body of text, wherein I have minimized scholarly language for the benefit of apprehension. More comprehensive issues of potential interest to those who are ready to become involved in additional (scientifically based) background information are summarized, in a condensed form, in sidebars. The interested and scientifically trained reader will also find citations of the original literature, introducing more complex issues. The scholarly reader will also find more profound astronomical and chemical information in two recent (text)books by the author: "Chemistry in Space – from Interstellar Matter to the Origin of Life", and "Bioinorganic Chemistry".

Dieter Rehder, Hamburg
Professor of Chemistry
Chemistry Department
University of Hamburg (Germany)
Tel: 004940428386087
E-mail: rehder@chemie.uni-hamburg.de
September 2014

Basic Measuring Categories and Units

Column density: N, in units of cm^{-2}, denotes the number of atoms/molecules/ions/particles (in the line of sight) per cm^2.

Density: ρ, in units of $g\ cm^{-3}$.

Distance: d, in units of meter (m); or in astronomical units AU, or in units of light years (ly), or in parallax of arcsecond (parsec, pc). 1 AU = $149.6 \cdot 10^9$ m, 1 ly = $9.46 \cdot 10^{15}$ m, 1 pc = 3.26 ly. See also Figure 2.2.

Frequency (of electromagnetic radiation): ν, in units of Hz. $\nu = c/\lambda$, where c is the speed of light, and λ the wave length.

Mass: m, in units of gram (g). Planets and stars in units of the mass of Earth = $m_{\oplus}$, the mass of Jupiter = m_J, and the mass of the Sun = $m_{\odot}$.

Number density: n, in units of cm^{-3}, denotes the number of atoms/molecules/ions/particles per cm^3.

Pressure: in units of Pascal, Pa. 1 Pa = 1 kg m^{-1} s^{-2}. 1 atm $\approx 1.013 \cdot 10^5$ Pa; 1 bar = 10^5 Pa.

Radiation: gamma (γ), X-ray, elemental particles: (i) Absorbed dose of radiation, in units of Gray (Gy), is the radiation energy (in Joule, J) per kg of absorbing matter; 1 Gy = 1 J kg^{-1}. (ii) Dose equivalent radiation for tissue, in units of Sievert (Sv); 1 Sv = 1 Gy $\times f$ J kg^{-1}, where f is a dimensionless weighing factor depending on the nature of the radiation. f = 1 (γ rays, electrons, low energy neutrons), f = 2 (protons, pions), f = 20 (α particles), f = 100 (high energy neutrons).

Speed: $v = d/t$ in units of m s^{-1}; d is the distance in meters (m), and t the time in seconds (s). Speed of light: $c = 299.8 \cdot 10^6$ m s^{-1} (in vacuum)

Temperature: T, in Kelvin (K, absolute temperature) or °C; 0 °C = 273.15 K. Conversion to Fahrenheit: °F = 9/5(°C) + 32

Time: t, in units of seconds (s), hours (h), days (d) or years (y). 1 y = 31 556 926 s. 1 My (megayear) = 10^6 y, 1 Gy (or Ga, gigayear) = 10^9 y.

Wave length (of electromagnetic waves): λ, in units of nanometer nm or Ångström Å. 1 Å = 0.1 nm.

INDEXING OF STARS

The following abbreviations, used in the present book, are commonly employed for the classification of stars and their planets. Planets are indicated by lower case letters, starting with the letter "b" in the sequence of their discovery; double and triple stars are distinguished by using capital letters (A, B etc.).

CoRoT	*Co*nvection *Ro*tation and planetary *T*ransits
CW	*C*epheid variables, named after the prototype *W* Virginis
GD	*G*iclas *D*warf
Gliese (Gl) and GJ	*Gliese* and (more recent) *G*liese and *J*areiß catalogue
HAT-P	*H*ungarian *A*utomated *T*elescope network *P*roject
HD	*H*enry *D*raper catalogue for star designation
HR	*H*arvard *R*evised number
IRAS	*I*nfrared *A*stronomical *S*atellite
Kepler	Exoplanets identified in the frame of the *Kepler* mission
KOI	*K*epler *O*bject of *I*nterest
LHS	*L*uyten *H*alf *S*econd
Luhmann	Kevin *Luhmann* (an astronomer)
2M	*T*wo *M*icron all sky survey
OGLE	*O*ptical *G*ravitational *L*ensing *E*xperiment
WASP	*W*ide *A*rea *S*earch for *P*lanets
WISE	*W*ide-field *I*nfrared *S*urvey *E*xplorer

TYPES OF STARS

(See also sidebar 2.2 for the classification of stars)

AGB Stars: AGB stands for *A*symptotic *G*iant *B*ranch, and relates to red giant stars formed as a main sequence stars (cf. sidebar 2.2) with $m > 4m_\odot$ developing from hydrogen fusion via helium fusion to carbon burning, thus forming elements beyond iron, and even molecular species.

Black holes: Black holes are remnants of supernova explosions of stars commonly with masses $m > 20m_\odot$ (up to a few million $m_\odot$). Black holes are massive to the extent that photons can no longer escape their gravitational field.

Brown dwarfs: 'Brown dwarf' is a collective term for lower mass stars ($m < 0.075m_\odot$) of low surface temperature (< 2300 K) and luminosity (late M stars and less). Core temperatures are too low for hydrogen burning. For a more distinct differentiation of 'brown dwarfs', see Table 2.2.

Carbon stars are particularly carbon-rich old stars, a subclass of the asymptotic giant branch stars.

Mira variables (named after the prototype Mira) are pulsating variable red giant stars in the very late stage of stellar evolution, where they develop into planetary nebulae and finally into white dwarfs. A subclass are the Cepheid variables.

Neutron stars: Remnants of type II supernovae with the following characteristics: $m \sim 1.4\ m_\odot$, $r \sim 10$ km. In the dense core of neutron stars, protons and electrons have been merged to neutrons and escaping neutrinos.

Nova: A nova forms in the course of a nuclear explosion (on the surface) of a white dwarf that accretes hydrogen from a companion star in a binary system. Novae can be precursors to type Ia supernovae.

Planetary nebula: Represents a late stage of the evolution of a star beyond the red giant state. Eponymous for planetary nebulae are characteristic emission nebulae expanding from the star into space.

Pulsars: A pulsar (pulsating star) is a neutron star that emits electromagnetic radiation in pulses. The electromagnetic radiation is a consequence of an inclination of the magnetic axis with respect to the axis of rotation. Typically, pulses are in the millisecond to the second regime.

Red dwarfs: A term often used for early M type stars ($m \approx 0.6$ to $0.08\ m_\odot$).

Red giants: A general term for luminous giant stars in an early or $-$ more commonly $-$ late phase of stellar evolution. The surface temperature is comparatively low (3000-4000 K). In the late phase, when the core hydrogen has been used up, hydrogen fusion proceeds in the stellar shell.

Supernovae: They are the final stage in the development of a massive star or a double star system. This final stage is characterized by an explosive disintegration of the star, accompanied by a brief increase of its luminosity by several orders of magnitude. Sub-categories: *Type Ia*: A merger of a white dwarf and any other star. Spectral characteristics: Si^+. *Type Ib:* Core-collapse of a massive star. Characteristics: He; no Si^+. *Type Ic* characteristics: neither H nor He; broad absorption lines for C, O, Si, and Mg. *Type II:* Core-collapse of stars with 8-20 (up to 50) $m_\odot$. Characteristics: distinct from type I by the presence of hydrogen lines in the spectra.

T-Tauri variables: These variables are very young stars, arising from the development of a protostar in a collapsing dark interstellar cloud. T-Tauri variables have a comparatively low mass ($m < 2m_\odot$), but are highly luminous and optically variable.

White dwarfs: Particularly dense stellar remnant of a developed, approximately Sun-like star; $m \sim m_\odot$, $r \sim r_\oplus$, $\rho \approx 10^6$ g cm^{-3}.

Wolf-Rayet stars: Hot and massive stars ($> 20\ m_\odot$); the progenitors to supernovae type Ib and Ic, shedding their mass with fervid stellar winds.

Chapter 1

INTRODUCTION

We are pretty well acquainted with our Solar System, its planets (Mercury, Venus, Earth, Mars, Jupiter, Saturn, Uranus and Neptune), dwarf planets (such as Pluto and Ceres), asteroids and comets. Just two of these planets, Earth and Mars, are in the "habitable zone", i.e., at a distance from the Sun where surface and subsurface temperatures allow, at least basically, for liquid water and thus for the presence of life in a form familiar to us. Life has so far only been detected on our home planet; whether or not primitive life forms are present on Mars, or have been present on the early Mars, is presently being investigated.

In the past two decades, several thousand exoplanets have been discovered, i.e., planets circling other stars or, in a few distinct cases, even unbound free-floating planets. As of November 2014, 1849 of these exoplanets in 1137 planetary systems have been confirmed, mainly in the frame of the Kepler mission. Extrapolation suggests that our Milky Way galaxy should harbor some 10 billion planets, with perhaps 10% of these "Earth-like", i.e., circling their sun in the habitable zone, and with more or less the size and average density of Earth and thus physical and geological features resembling those of Earth.

Did life evolve on these siblings of our home planet as well? And what are the conditions – and probabilities – for life to evolve and to develop? On Earth, life is not restricted to "common" conditions – 'common' meaning mild temperatures in the oxygenic environment we are familiar with. Rather, when life commenced on our planet some 3.6 billion years ago, oxygen was hardly available and the radiation flux of our Sun was just about two-thirds of that of today. Many of the primordial microbial organisms developed and dwelled in "hostile" environments, and still thrive today in habitats such as hot smokers in deep sea areas, high and frosty mountain areas subjected to intense UV and γ radiation, or in alkaline or highly acidic media with extreme salt concentrations and temperatures up to 120 °C. Are there after all conceivable scenarios for life to have evolved on exoplanets as well and if so: have there been parallel lines for the development of life, or did life develop just once, and then become "dispersed" through space from its singular, original source of origin to other planetary systems?

The present treatise will first provide insight into the buildup of our Solar System which, as far as we can presently judge, appears to be atypical of planetary systems in the Milky Way galaxy. Exoplanets are then classified, and detection methods are briefly addressed in Chapter 2. Chapter 3 conveys an overview on the formation – and subsequent development – of planetary systems out of collapsing interstellar clouds. In this context, some background chemistry in dark interstellar clouds is surveyed and thus molecular compounds – start-ups for more complex molecules (including those required for the generation of life molecules) – that became integrated into the newly formed planetary bodies. This issue is then broadened and specified in Chapters 4 and 5; Chapter 4 deals with geological and atmospheric features of exoplanets, and is further concerned with the question of how we can extract relevant information from objects (meteorites in particular) stemming from regions beyond the habitable zone, while Chapter 5 defines "Life". Furthermore in Chapter 5, microorganisms flourishing in extreme habitats will be dealt with, as well as putative life forms based on a non-carbon world, and independent of water.

The issues "generation of life molecules" and "initiation of life from inanimate (macro)molecules" are then retrieved in Chapter 6. Chapter 7 enlarges upon the notion of "habitability", as well as the disposability and requirement of water. Saturn's moon Titan is introduced as a potential model for the development of primitive life beyond Earth and our Solar System. In this context, the notion "panspermia" will be dealt with. Panspermia is a somewhat hypothetical view on how sperms of life and/or molecules containing information relevant to the generation of life might have been (and are still being) distributed through space, and hence a view that approaches the questions "is life unique to our planet?" and "is our planet the genuine source of life?"

Chapter 2

PLANETS AND EXOPLANETS: METHODS OF DETECTION

Planets are celestial bodies orbiting a central star (or, eventually, a binary or ternary system), where the term "star" encompasses the main sequence of stars in the Hertzsprung-Russel diagram, including dwarf stars but, to some extent, also (developed) stars beyond the main sequence, such as red giants and white dwarfs.

As far as our Solar System is concerned, the term "planet" has been stringently defined by the International Astronomical Union. In extrasolar systems, this definition is less strictly applied. Thus, along with planets circling their sun, there are also unbound planets (that possibly "escaped" their place of birth in an early stage of the development of their solar system). Also, in some cases, the distinction between a system consisting of a sun and its giant planet vs. a double star system is not clear-cut; in this context, a planet is then defined as a celestial object that does not produce energy by itself by way of nuclear fusion. The maximum mass limit for a planet is thus commonly quoted as thirteen Jupiter masses. Beyond this limit we are dealing with brown (sub)dwarfs.

A large number of detected extrasolar planets move on orbits pretty close to their sun, clearly within Mercury's orbit in our system, and thus at time intervals encompassing just a few days. At least in part, this discrepancy with our Solar System is due to the detection methods; detection is, in many cases, more straight as a planet is close to its star. On the other hand, a couple of giant planets orbit their sun at distances far beyond Neptune's orbit, in a region that, for our Solar System, is known as Kuiper belt, provoking the question of whether their actual nature more likely is a binary star.

In this chapter, the following questions will be addressed: "what is an exoplanet", "how can exoplanets be classified", and "how are exoplanets detected". In this context, we will start with a brief description of the Solar System, focusing on the key properties of the Solar System's planets, selected dwarf planets and moons. We will further introduce units (such as distance and mass) commonly used in our system and also employed in extrasolar systems, and provide a brief excursus on atom design and radioactivity, hence properties that are also of interest in the context of distinguishing a "sun" from a "planet".

2.1. PLANETS: DEFINITION, AND DISTINCTION FROM BROWN (SUB)DWARFS

In 2006, the International Astronomical Union (IAU), on the occasion of a meeting in Prague, redefined the term planet – not without controversy – by fixing four criteria in the following way:

1. A planet is a body in orbit around the Sun;
2. A planet has sufficient mass for its self-gravity to assume a hydrostatic equilibrium and hence a nearly globular shape;
3. A planet has cleared the neighborhood around its orbit from other objects;
4. The body under consideration is not a satellite.

In some respects, criterion (3) lacks clarity: "cleared the neighborhood around its orbit" may be interpreted in terms of (i) "cleared from debris", namely objects ranging from dust particles to irregular rocky fragments; and (ii) "the orbits of these fragments are dictated by the central body (the planet)". The applicability of point (i) is questionable in the light of the 'debris' that at least temporarily orbits other planets, such as Saturn and Uranus. As far as point (ii) is concerned, Pluto loses the status of a planet, since the motion of the Pluto system (Pluto and Charon, plus the four minor moons) is dictated by Pluto's massive neighbor Neptune: Pluto is locked to Neptune in a 2:3 resonance, meaning that the Pluto system circles the Sun twice during three circuits of Neptune.

In any case, Pluto has officially lost its status as a planet and is now categorized as a "dwarf planet" – along with an increasing number of other trans-Neptunian, so-called Kuiper belt objects, among these Haumea,

Makemake (pronounced mɑːkiːˈmɑːki), Eris (slightly larger than Pluto and hence the largest Kuiper belt object), Sedna and Quaoar; and the asteroid Ceres. Consequently, the popular mnemonic helping to remember the sequence of planets, "*My* *V*ery *E*legant *M*other *J*ust *S*erved *U*s *N*ine *P*izzas", might be varied to "*My* *V*ery *E*legant *M*other *J*ust *S*erved *U*s *N*achos", where the capital letters in bold stand for Mercury (the planet closest to the Sun), Venus, Earth, Mars, Jupiter, Saturn, Neptune (the planet farthest from the Sun), and Pluto the dwarf planet. Figure 2.1 provides a notion of the size of the planets and selected dwarf planets of the Solar System. In between the orbits of Mars and Jupiter there are the asteroids (forming the asteroid belt[1], stretching between 2.1 to 3.3 AU), with their largest member, Ceres, fulfilling – as Pluto – criteria no. (1) and (2) only, and thus also classified as a *dwarf planet*.

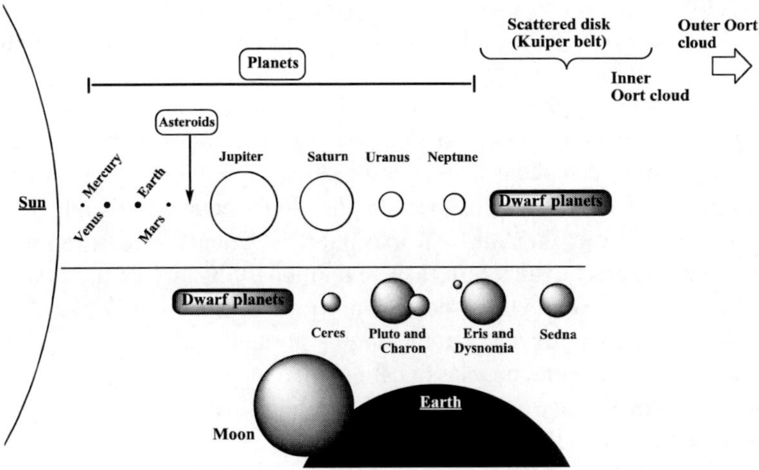

Figure 2.1. The Solar System: The size of planets in relation to the Sun. The upper part shows the relative size and distances of the planets, the lower part the size of selected dwarf planets, Earth's moon, Pluto's moon Charon, and Eris' moon Dysnomia in relation to Earth (lower part). Distances from the Sun are not to scale. The Kuiper belt extends from Pluto to about 70 AU, the inner Oort cloud (the home of the most distant dwarf planets Sedna and 2012 VP_{113}) to ca. 1,000 AU, and the outer Oort cloud (from where most of the comets originate) to about 100,000 AU.

---

[1] Occasionally, asteroids can develop a form of appearance reminiscent of comets. These active asteroids, also termed "main belt comets", represent a distinct class of solar system bodies in the asteroid belt that develop (temporary) dust tails due to morphological changes, e.g., as a result of high rotation rates.

There is evidence that our present-day planetary system has not always stayed in its current form. Rather, the giants Jupiter and Saturn are believed to have migrated through the Solar System some 4.5 to 3.9 billion years ago, thus also strongly influencing the inner planets. Such scenarios certainly have consequences also for extrasolar systems, where Jupiter-sized planets have been found in orbits pretty close to their sun. The most prominent of these migration scenarios in our system are the Nice Model and the Grand Track Model. According to the Grand Track Model, Jupiter and Saturn formed in the protoplanetary disk just outside the snowline (the limit beyond which temperatures are sufficiently low to enable the existence of water ice), and jointly migrated outward after they became locked into a 3:2 resonance (of their orbital periods). A similar scenario is described by the Nice Model (with a 2:1 resonance). This past motion coupling and reorganization within the early Solar System is also responsible for the assembly of Kuiper Belt and Scattered Belt objects, and their gravitational scattering to regions far beyond the outer asteroid belt, and the late heavy bombardment of the inner planets approximately 3.9 billion years ago.

Table 2.1 summarizes characteristics of the planets and selected dwarf planets of the Solar System, viewed here as a reference for extrasolar planetary systems. Typically, the planets revolve on slightly eccentric elliptical orbits only marginally inclined with respect to the Sun's equator, rotating in the direction as they orbit the Sun. An exception is Uranus with a retrograde rotation. These characteristics reflect the evolutionary history of the Solar (as well as extrasolar) systems to be dealt with in more detail in Chapter 3. The dwarf planets largely deviate from these "normal" orbit characteristics. Benchmarks for the common classification of exoplanets (vide infra) with respect to size and density are the rocky planet Earth with an iron-nickel core, the planet Neptune, and the gas giant Jupiter.

Neptune and Jupiter are commonly used − together with Earth − as a reference system for (the size and, to some extent, also the composition of) exoplanets. Neptune's atmosphere makes up about 5-10% of the planet's mass; the main atmospheric constituents are molecular hydrogen H_2, helium He, and methane CH_4. The planet's mantle is composed of a fluid mix of water H_2O, ammonia NH_3 and CH_4; in greater depths, hydrogen ions, a lattice of oxygen atoms, and elemental carbon (possibly diamond-like modifications) dominate. The core consists of iron-nickel and silicates. The predominant gases in Jupiter's atmosphere are H_2 and He. The structure and composition of the mantle and core are hardly known. The inner core is believed to be rocky, surrounded by a layer of liquid metallic hydrogen. The outer core layer again is H_2.

Table 2.1. Characteristics of the planets, as well as of selected dwarf planets and moons in the Solar System

	Equatorial radius relative to Earth $= 1^{c}$	Mass relative to Earth $= 1$	Mean density ($g\ cm^{-3}$)	Mean distance from Sun (AU^{d})/planet (km)	Orbital period (years/days)[e]	Orbit inclination to ecliptic[f] (°)	Eccentricity[g]
Mercury	0.383	0.055	5.427	0.387	0.241	7.005	0.206
Venus	0.95	0.815	5.243	0.723	0.615	3.396	0.007
Earth	1	1	5.515	1	1	0	0.017
Mars	0.533	0.107	3.934	1.524	1.881	1.850	0.093
Ceres[a]	0.00015	0.076	2.077	2.767	4.80	10.593	0.076
Jupiter	11.21	318	1.326	5.06	11.86	1.305	0.049
Saturn	9.45	95.2	0.687	9.61	29.46	2.485	0.056
Uranus	4.01	14.5	1.27	19.20	84.32	0.773	0.044
Neptune	3.88	17.2	1.638	30.10	164.79	1.768	0.011
Pluto[a]	0.18	0.0022	2.03	39.26	247.68	17.15	0.248
Eris[a]	0.18	0.0028	2.52	67.96	560.23	43.86	0.437
Sedna[a]	0.16	0.00017	~1	506.7	11 400	11.93	0.853
Moon[b]	0.273	0.0123	3.346	$3.84\cdot10^{5}$	27.3	5.145	0.055
Ganymede[b]	0.413	0.025	1.936	$1.07\cdot10^{6}$	7.15	-	0.0013
Titan[b]	0.404	0.0225	1.88	$1.22\cdot10^{6}$	15.95	-	0.029
Enceladus[b]	0.04	$1.8\cdot10^{-5}$	1.61	$2.38\cdot10^{5}$	1.37	-	0.005

[a]Dwarf planets. [b]Moons: Earth's Moon, Ganymede (Jupiter system; largest moon in the solar system), Titan (Saturn system; most massive moon, and the only moon with a dense atmosphere; section 7.3), Enceladus (Saturn system; featuring a subsurface ocean). [c]The radius of Earth at the equator is 6,352 km. [d]For the definition of the Astronomical Unit (AU), see Fig. 2.2. For moons, the mean distance (from the planet) is provided in km. [e]For planets and dwarf planets in years, for moons in days. [f]The ecliptic is the plane of the path of Earth around the Sun. [g]Ellipticity; can vary between >0 (circular = 0) and <1 (hyperbolic).

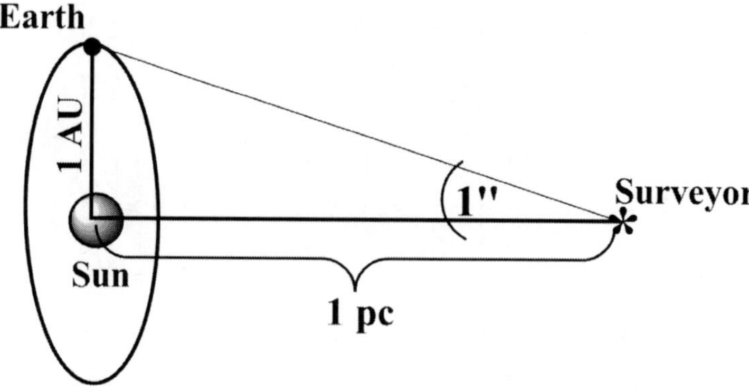

Figure 2.2. Definition of the astronomical unit (AU) and the parsec (parallax of arcsecond, pc): 1 AU is the mean distance Sun-Earth = $149.6 \cdot 10^6$ km; the mean relative position of the Sun is indicated; the ellipticity of Earth's orbit is grossly exaggerated. 1 pc is the distance at which 1 AU is viewed at an angle of 1 arcsecond (1"). The two distance parameters are related to each other through the tangent function: *tan*(1") = 1AU/1pc. 1 pc is equal to 3.26 light years (ly), i.e., the distance the light covers in vacuum within 1 second; 1 ly = 299.8×10^3 km.

The Kuiper belt, extending to approximately 50 AU from the Sun, is the innermost region of the Scattered Disk (extending well beyond 100 AU), a reservoir of icy dwarf planets commonly with highly eccentric orbits substantially inclined with respect to the ecliptic. At a distance of about 120 AU, influences through the solar wind on the one hand, and the interstellar medium on the other, balance. This outermost edge of the Sun's sphere (or heliosphere) is referred to as heliopause. The space craft Voyager 1, launched in 1977, passed this boundary in August 2012. With respect to gravitational influences, the outer limit of our Solar System is represented by the Oort cloud, an assemblage of planetesimals, consisting of ice and dust. The outer Oort cloud extends to a distance of approximately 1 light year (50,000 AU), and hence one-quarter of the distance towards the Sun's closest neighbor, Proxima Centauri, part of a triple system with α-Centauri. Occasional gravitational disturbance of Oort cloud objects can destabilize these planetesimals into highly elliptical or even hyperbolic paths around the Sun, where they appear as comets when passing by the Sun at a sufficiently close distance.

The strict definition of the term 'planet' provided above reflects the situation in our Solar System, and thus does not necessarily apply to exoplanets, i.e., planets circling fixed stars other than our Sun, or even free floating planets. Exoplanets orbiting suns beyond our Solar System can vary in size from sub-Mercuries to super-Jupiters, with an upper limit commonly appointed (by the IAU) 13 Jupiter masses, m_J. This applies to objects with the metallicity of the Sun, where "metallicity" refers to the ratio metal/hydrogen [Fe/H], and metal (Fe) stands for all elements beyond helium. For the Sun, [Fe/H] = 0 by definition.[2] The Sun [Fe/H] > 0, for those older than the Sun [Fe/H] < 0. Beyond a mass > 13 m_J, deuterium fusion starts. Nuclear fusion is a property attributed to stars, and hence objects with more than 13 m_J are no longer considered planets.

In deuterium fusion, also referred to as deuterium burning, a proton and a deuterium nucleus (proton + neutron) combine to form the helium isotope of atomic mass 3, Eq. (2.1), accompanied by γ radiation. For terms related to nuclear reactions, see also sidebar 2.1. Stellar objects with masses >13 m_J and <90 m_J are usually subsumed as brown dwarfs (for a more detailed classification see below), and thus as stars where the core temperature is too low to allow for hydrogen fusion. Hydrogen fusion is the main process of energy supply in stars such as our Sun. The net reaction is represented by Eq. (2.2); along with helium, positrons and neutrinos are generated. A typical brown dwarf is Gliese 229B in the constellation Lepus. Gliese 229B comprises ca. 35 m_J, but is considerably denser than Jupiter, with a diameter about 10% less than Jupiter. Brown dwarfs with masses > 65 m_J recur, in addition to deuterium burning, to lithium burning. Lithium burning, Eq. (2.3), is the fusion of lithium-7 and protons to form the common helium isotope of mass 4. When the star's mass exceeds ca. 80 m_J, hydrogen burning starts, and these stellar objects are then no longer accommodated within the family of brown dwarfs.

$$\ce{^{1}_{1}H} + \ce{^{2}_{1}H} \longrightarrow \ce{^{3}_{2}He} + \gamma \tag{2.1}$$

$$4\,\ce{^{1}_{1}H} \longrightarrow \ce{^{4}_{2}He} + 2\,\ce{^{0}_{1}e^{+}} + 2\nu \tag{2.2}$$

$$\ce{^{7}_{3}Li} + \ce{^{1}_{1}H} \longrightarrow 2\,\ce{^{4}_{2}He} \tag{2.3}$$

***

---

[2] The explicit expression for metallicity is $[Fe/H] = \log_{10}(n_{Fe}/n_{H})_{star} - \log_{10}(n_{Fe}/n_{H})_{Sun}$, where n is the number density (for the definition of n see footnote 3).

Sidebar 2.1. Buildup and decay of atoms

Atoms consist of a nucleus and a shell (electrons); see the table below for characteristics of selected elemental particles. In a simplistic atomic model, the electrons (e⁻) orbit the nucleus on shells with a defined distance from the nucleus. In the more realistic atomic orbital model, electrons are no longer particles, but are described in terms of waves, or the probability of locating the electron in a specific region – the orbital – around the nucleus; see also sidebar 5.3. The charge of the electron is, by definition, -1.

The nucleus of an atom contains protons and neutrons (with the exception of the lightest hydrogen isotope, protium, where there is just a single proton). Protons (p) have a charge of $+1$, while the charge of neutrons (n) is 0. In a neutral atom, the number of protons equals the number of electrons. The mass of an atom is dominated by the sum of protons and neutrons. Charged atoms, or ions, possess an excess of electron(s) (anions) or a deficit of electron(s) (cations). For the lighter elements, the number of protons in a nucleus is about the same as the number of neutrons; gross imbalances result in instability and thus radioactive decay (vide infra). The number of protons identifies the element. Nuclei with a defined number of protons but differing numbers of neutrons are termed isotopes; isotopes are thus variants of the same element. Examples are protium, deuterium and tritium, the three isotopes of hydrogen, with none, one and two neutrons along with the single proton.

Some properties of common elementary particles

	Symbol	Charge (in elementary charge units)	Rest mass (in elementary mass units)	Spin
Proton	p	$+1$	1.00728	1/2
Neutron	n^a	0	1.00866	1/2
Electron	e^-, β^-	-1	0.0005486	1/2
Positron	e^+, β^+	$+1$	0.0005486	1/2
Antineutrino and Neutrino	$\bar{v}$ and v	0	0	1/2

^{a} Neutrons are unstable; their half-life is 885.7 s.

To identify a nucleus, the number of protons is added as a lower left index to the element symbol, the sum of protons and neutrons as an upper left index; examples are provided in Eqs. (2.1)–(2.3). A charge of an ion is indicated by an upper right index, but commonly omitted in equations describing nuclear reactions.

Protons and neutrons in turn consist of quarks: protons are comprised of two quarks (or up-quarks) and one antiquark (or down-quark), neutrons of one quark and two antiquarks. Quarks hold together via gluons (weak interaction), while protons and neutrons have strong interaction via bosons. The strong interaction between one quark and one antiquark produces a short-lived (micro- to milliseconds) meson.

In a radioactive nucleus, a neutron or (less frequently) a proton decomposes, and a new nucleus – the daughter nucleus – forms. The main variants of radioactive decay for lighter nuclei with a proton/neutron imbalance are β (electron) and the β^+ (positron) decay, with the following underlying elemental processes:

$$\beta^- \text{ decay: } {}^1_0n \longrightarrow {}^1_1p + {}^0_{-1}\beta^- + {}^0_0\bar{\nu} \qquad \text{Example: } {}^{14}_6C \longrightarrow {}^{14}_7N + {}^0_{-1}\beta^- + \bar{\nu}$$

$$\beta^+ \text{decay: } {}^1_1p \longrightarrow {}^1_0n + {}^0_1\beta^+ + {}^0_0\nu \qquad \text{Example: } {}^{17}_9F \longrightarrow {}^{17}_8O + {}^0_1\beta^+ + \nu$$

Complementary to the β^+ decay is electron capture, i.e., the capture of an electron by a proton to form a neutron:

$$\text{electron capture: } {}^1_1p + {}^0_{-1}\beta^- \longrightarrow {}^1_0n + {}^0_0\bar{\nu} \qquad \text{Example: } {}^{40}_{19}K + {}^0_{-1}\beta^- \longrightarrow {}^{40}_{18}Ar + \bar{\nu}$$

The decay is commonly accompanied by γ radiation, an electromagnetic radiation in the extreme short wave (high energy) regime, i.e., the daughter nucleus accrues in an excited state and rapidly stabilizes by passing to its ground state, emitting the energy difference as γ rays.

Electron capture and β^+ decay are also the source for the formation of most of the iron in the Universe. The reaction starts from ^{56}Ni and proceeds via ^{56}Co:

$$\begin{array}{l} {}^{56}_{28}Ni + {}^0_{-1}\beta^- \longrightarrow {}^{56}_{27}Co + \nu + \gamma \qquad (t_{1/2} = 6 \text{ d}) \\[1em] \qquad\qquad\qquad\quad \Big\downarrow {}^{+{}^0_{-1}\beta^-} \\[0.5em] \qquad\qquad\qquad\qquad \longrightarrow {}^{56}_{26}Fe + \nu + \gamma \qquad\qquad (81\,\%,\ t_{1/2} = 77 \text{ d}) \\[1em] \qquad\qquad\qquad\qquad \longrightarrow {}^{56}_{26}Fe + {}^0_1\beta^+ + \nu + \gamma \quad (19\,\%,\ t_{1/2} = 111 \text{ d}) \end{array}$$

The term $t_{1/2}$ indicates the half-life; here provided in days. The short-lived radionuclide ^{56}Ni can be generated in a cascade of α captures (α particles are ^{4}He nuclei) starting from ^{4}He or ^{12}C and proceeding via ^{28}Si:

$$^{28}_{14}\mathrm{Si} + {}^4_2\mathrm{He} \longrightarrow \longrightarrow \longrightarrow {}^{56}_{28}\mathrm{Ni} + \gamma$$

***

Despite their name, brown dwarfs are not necessarily brown in appearance. Early M-type brown dwarfs are therefore also termed red dwarfs. Depending on the surface temperature – ranging from approximately 2300 to 300 K – they can appear orange-red and magenta. In the classical classification of stars based on their surface temperature and luminosity, "brown" dwarfs are at the low luminosity/low temperature margin, the so-called late M stars, in a series starting with particularly hot and hence bright stars, the O stars, followed by B, A, F, G, K and M stars, a sequence for which the mnemonic *"Oh Be A Fine Girl/Guy Kiss Me"* holds. This sequence is further subdivided in ten subclasses (0 to 9) per main class. As an example, our Sun with an effective surface temperature of 5780 K and – by definition – a luminosity of 1, is a G2 star. For a more detailed account on the classification of stars, see sidebar 2.2. Late M stars, which are insufficiently hot to allow the nuclear fusion of hydrogen (to form helium), are sometimes considered "incomplete" stars or "substellar" objects. The classical sequence of stars has now been extended, at the low luminosity/low temperature end, to L, T and – provisionally – Y stars (Cushing 2011; Dupuy 2013).

Table 2.2. Classification of brown dwarfs

Type	Surface temperature (K)	Mass in solar masses $m_\odot$	Spectral features dominated by	Prototype[b]
M	2300 (upper limit)	<0.075[a]	TiO, VO	LHS 2065
L	2000-1300	<0.065	FeH, CrH, MgH, CaH; Na, K, Cs, Rb; H_2O, CO	GD 165B
T	1300-700	0.06-0.02	Na, K; CH_4, (H_2O, CO)	GJ 229B
Y	<600 (down to ≈250)	0.02-0.005	H_2O, CH_4, NH_3 (?)	WISE 1828+2650

[a] The range of M dwarfs can extend to 0.6 $m_\odot$; these more massive and hotter M dwarfs are commonly termed "red dwarfs". [b] The capital letter "B" in GD 165B and GJ 229B indicates the second (less massive) component in these binary stellar systems. For the acronyms LHS, GD, GJ and WISE see "Indexing of stars".

Table 2.2 provides a classification for brown dwarfs according to surface temperature, mass and typical spectral features. Objects beyond Y stars are exoplanets, and at least several Y stars are at the boundary of exoplanets and therefore occasionally also classified as free-floating, or rogue, exoplanets.

The most recent example of a super-cool brown Y dwarf is WISE 0855-0714, only about seven light years from the Sun (and hence ranking 4th in proximity to the Sun), with a temperature of 260-225 K, and a mass of 3-10 m_J (Luhman 2014).

Brown dwarfs can harbor exoplanets. An example is 2M 1207 (with a mass of 26 m_J) in the constellation Centaurus, and its planet 2M 1207-b. Exoplanets are marked by lower case letters (b, c, d etc., in the sequence of their discovery), following the identification of their sun. In contrast, the *stars* in double and multi-star systems are differentiated by capital letters A, B etc. 2M 1207-b (3-10 m_J) and is their sun at a distance of 40.6 AU and hence are pretty far outside the range common for the large majority of exoplanets: Most exoplanets orbit their sun in the range observed for our Solar System, or even much closer than Mercury, the innermost planet in our system, in accordance with the commonly accepted model for the genesis of a sun plus the manifold of the sun's planets, as described in Chapter 3. Even more remote from its sun, at a distance of 650 AU, is the exoplanet HD 106906-b, with about 11 mJ (Baily 2014).

***

Sidebar 2.2. Classification of stars

Depending on their surface temperature and luminosity, stars are classified into O, B, A, F, G, K, and M spectral classes, where O stars are the hottest and M stars the coldest representatives. The benchmark for luminosity L is the Sun ($L = 1$). The luminosity of a star is defined as the amount of energy a star radiates per second, and is related to the temperature T and the radius r of the star: $L = 4\pi k r^2 T^4$, where k is the Boltzmann constant.

The overall situation is represented in the Hertzsprung-Russel (HR) diagram. Within the classifications O, B, A etc., sub-classes are defined, such as G0 to G9, where '0' is at the high-temperature and '9' at the low-temperature margin. Our Sun, with a surface temperature of 5780 K, is a G2 star. An additional descriptor, indicating the category the star belongs to, is sometimes added, e.g., – in the case of the Sun – G2V, where the Roman numeral "V" denotes the Sun's affiliation to the main sequence. Typical characteristics of main sequence stars are compiled in the table below.

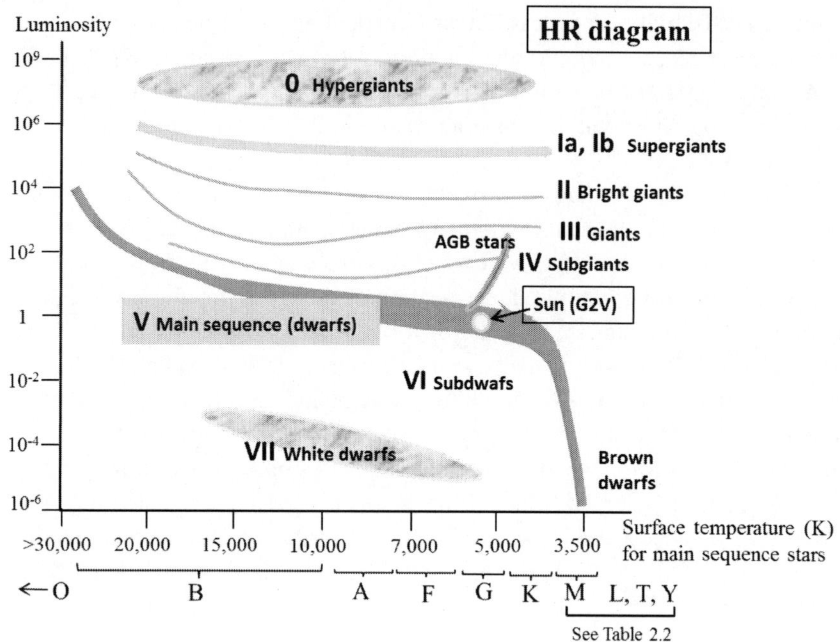

Cosmic abundances range from 10^{-5} % (O stars) to 3.5 % (G stars) and 80 % (M stars). AGB (the about vertical line in the HR diagram close to the Sun) stands for Asymptotic Giant Branch stars, a region that is populated by stars of low to medium mass in a late state of their evolution.

In addition to the spectral class, the "color index" of a star is sometimes enlisted. The color index stands for the difference of the star's *magnitude* at short wave length (e.g., blue, *B*) and long wave lengths (e.g., green-yellow, *V* [for visible]), hence *B-V*. The "absolute magnitude" m_{abs} of a star is defined by $m_{abs} = -5/2 \times \log_{10}(L/L_{\odot}) + 6$. Hot stars such as stars belonging to the B and A categories have major contributions in the short wave length range, while for cooler stars, G, K and M stars in particular, the long wave length components predominate.

Selected characteristics of main sequence stars

Spectral class	Color	Surface temp. (K)	Luminosity relative $L_\odot$	Radius relative $r_\odot$	Mass relative $m_\odot$	$\varnothing$ Life time in 10^6 a	Prominent emission lines
O	Blue	>30,000	10^5	>7	>17	2–8	He, He$^+$, C^{2+}, N^{2+}, O^{2+}, Si^{4+}
B	Blue-white	30,000–9,500	$3 \cdot 10^4$–25	7–2.4	16–2.9	9–200	H, He, Mg$^+$, Si$^+$
A	White	9,500–7,200	25–5	2.4–1.5	2.9–1.6	300–10^3	H, Mg$^+$, Si$^+$, Fe$^+$
F	Yellow-white	7,200–6,000	5–1.5	1.5–1.15	1.6–1.04	2–7·10^3	H, Ca$^+$, Fe, Cr
G	Yellow	6,000–5,200	1.5–0.6	1.15–0.9	1.04–0.8	8·10^3–2·10^4	Ca$^+$, M, M^{+b}
K	Orange	5,200–3,800	0.6–0.08	0.9–0.6	0.8–0.5	2·10^4–7·10^4	Si, Fe, Mn
M^a	Red	<3,800	<0.08	<0.6	<0.6	7·10^4–10^6	M^b, TiO, VO

^{a}For a more detailed treatment of cooler M and sub-M dwarfs, see Table 2.2 in the main text. ^{b}M in this specific column represents other metals.

Spectral class:	B0	A0	F0	G0	K0	M0
B-V	-0.30	-0.00	+030	+0.58	+0.81	+1.40

***

2.2. CLASSIFICATION OF EXOPLANETS

Commonly, a planet is considered a celestial body orbiting its parent star. To be more precise, the star and its planet(s) move around a common center of gravity – an important issue in the context of the detection of exoplanets, section 2.4. The mass of the parent star generally exceeds that of its planetary companion(s) in such a way that the center of gravity is located within the star. Our solar system provides an example of a situation where this is *not* the case; namely, the Pluto-Charon system (cf. Figure 2.1): Pluto's largest moon Charon, at a distance of 19,600 km from the center of Pluto, has 11.6% Pluto's mass; the barycenter (center of mass) of the Pluto-Charon system thus lies outside either body.

"An infinity of worlds comparable to our own" had already been postulated by the Italian priest and universal scholar Giordano Bruno by the end of the 16th century (see also Preface). The first three exoplanets orbiting the pulsar PSR B1257+12 in the constellation Virgo have been discovered in 1992. These "second generation planets" have likely formed from remnants of the supernova that preceded the pulsar, and are thus hardly comparable with the "first generation planets" emerging from a protosolar system. The first well-documented discovery of a first generation exoplanet, 51 Peg-b, goes back to 1995 (Major 1995): 51 Peg, in the constellation Pegasus, is a G2 and thus a Sun-like star at a distance of 60 ly. The planet 51 Peg-b orbits its sun at a distance of only 0.05 AU; this compares to 0.39 AU for Mercury, the innermost planet in our Solar System. The mass of 51 Peg-b mass is about half that of Jupiter. Presently (as of November 2014), 1849 exoplanets belonging to 1160 planetary systems have been confirmed, with more than 2000 additional systems, mostly detected by the Kepler mission space telescope, awaiting confirmation. An actualized list of confirmed planets can be found at http://exoplanet.eu/catalog.php.

The Kepler mission - a space observatory named in honor of the mathematician and astronomer Johannes Kepler (1571-1630), famous, *inter alia*, for his laws of planetary motion - was launched in March 2009 and collected data until May 2013, when technical problems disabled the collection of further data. To that date, the observatory had monitored over 145,000 stars by assembling characteristics related to periodic dimming of the stars due to transiting potential planets. Extrapolating the present stand of knowledge to the one hundred billion (10^{11}) stars constituting our local galaxy, the Milky Way Galaxy, and assuming that, on average, each of these stars harbors one planet, there are 10^{11} planets around in our galaxy, with about one-tenth orbiting Sun-like stars. According to recent estimates (Petigura 2013), about 6% of the exoplanets are Earth-sized *and* orbit their stars in the habitable zone, at distances of 0.5 to 2.0 AU and with orbital periods of 200-400 days. "Habitable zone" is defined as that area where the planet receives between one and four times the stellar intensity as Earth, and where liquid water to sustain life as we know it can exist. "Earth-size" thus does not necessarily mean habitable. We will extemporize on the topic "habitability" in more detail in Chapter 5.

Based on the size (and mass) of the exoplanets, one roughly distinguishes between the following categories of exoplanets: (1) Mercury- to Mars-sized, (2) Earth-sized, (3) super-Earths and mini-Neptunes; (4) Neptune-sized, (5) Jupiter-sized (hot Jupiters), and (6) super-Jupiters (up to 13 m_J; vide supra). Of the confirmed exoplanets, about 1% belongs to categories (1) plus (2). A sub-Mercury-sized exoplanet, Kepler 37-b, has recently been detected (Barclay 2013). Presently, about 11% of the confirmed exoplanets are super-Earths (with masses between 2 and 10 $m_\oplus$), 15% are Neptune-sized, and the remaining ca. 70% are hot Jupiters and super-Jupiters. Certainly, this classification is somewhat arbitrary, and the smaller objects are grossly under-represented, simply because they are less easily detected or remain undetected at all. As already insinuated above, extrapolation of the data so far available suggests that about one in five stars could have an Earth-sized planet in the habitable zone (Petigura 2013). With an improvement of detection methods, the abundance of exoplanets is likely to change in favor of Earth-sized planets and super-Earths/mini-Neptunes. Table 2.3 provides a sampling of well-characterized exoplanets, all of which – in this sampling – circuit their sun on very close orbits, actually inside the orbit of Mercury in the Solar System.

Table 2.3. Examples of exoplanets

Designation[a]	Mass m[b]	Radius r[c]	Semi-major axis (AU)	Orbital period (d)	Mass of sun, $m_\odot$	Metalicity of sun[d]
Earths and super-Earths						
KOI 115-d	2.0 $m_\oplus$	1.9 $r_\oplus$	0.077	0.91	1.28	-0.22
Kepler 78-b	1.7 $m_\oplus$	1.2 $r_\oplus$	0.01	0.355	0.84	-0.08
Kepler 51-b	2.1 $m_\oplus$	6.9 $r_\oplus$	0.25	45.15	1	-
Kepler 138-c	1.0 $m_\oplus$	1.5 $r_\oplus$	-	23.0	0.57	-
Hot Jupiters						
CoRoT 1-b	1.03 m_J	1.49 r_J	0.025	1.51	0.95	0.06
HD189733-b	1.14 m_J	1.14 r_J	0.03	2.22	0.8	-0.03
Kepler 40-b	2.2 m_J	1.17 r_J	0.08	6.87	1.48	0.1
Kepler 44-b	1.02 m_J	1.24 r_J	0.045	3.02	1.19	0.26
Super-Jupiters						
CoRoT 27-b	10.4 m_J	1.01 r_J	0.048	3.57	1.05	0.1
Kepler 75-b	9.9 m_J	1.03 r_J	0.08	8.88	0.88	-0.07
WASP 14-b	7.34 m_J	1.28 r_J	0.036	2.24	1.21	0

[a]Cf. "**Indexing of stars**" for abbreviations. [b]In units of $m_\oplus$ (mass of Earth) or m_J (mass of Jupiter). [c]In units of $r_\oplus$ or r_J. [d]A measure for the ratio metal/hydrogen, where 'metal' stands for all elements heavier than helium; see also footnote 2.

Figure 2.3 is a depiction of the situation in an extrasolar multi-planet system in an early phase of its formation (Marois 2010), with the four super-Jupiters orbiting their central star HR 8799 in the constellation of Pegasus at distances that compare to the range spanned between Saturn and the Kuiper belt in our system. The Sun-like star HD 10180 in the constellation of Hydrus harbors the so-far largest number of exoplanets (Tuomi 2012). To date, six planets, HD 10180-c to HD 10180-h, have been confirmed, with masses comparable to Uranus or Neptune, and distances from their sun ranging from 0.06 to 3.4 AU.

Planetary systems commonly form as an interstellar dark cloud – a cloud containing essentially molecular hydrogen and up to 1% of dust particles – collapses, typically as a result of a shock wave generated by a supernova explosion (see Chapter 3). The collapse produces a disk-like, rotating pre-planetary nebula from which a protosun and a debris belt form, the latter providing the material for planetesimals, most of which finally aggregate to form planets. Sun and planet can also form simultaneously. An example is the system HD 106906A/HD 106906-b in the Lower Centaurus Crux constellation. 106906-A is a young (ca. 13 My old) F5 star (and thus about 1.5 times as bright as our Sun) with a debris belt extending from 15 to 120 AU. The star is orbited by a companion of ca. 13 m_J at a mean distance of 650 AU (Bailey 2014).

As noted, the definition employed for a planet in our Solar System provided in section 2.1 is not strictly applied to exoplanets. Thus, some planets, referred to as "rogue planets", are ejected from the system, leaving the system on hyperbolic orbits and thus escaping into free space. These unbound, free-floating planets straying across the galaxy, also termed "Steppenwolf planets" (Abbot 2011), may harbor a subsurface ocean, similar to Saturn's moons Enceladus and Titan. They are of some interest in the context of extraterrestrial life and, in particular, of "panspermia", the potential distribution of life throughout the Universe. For a critical account on panspermia, see section 7.4.

Whether or not a free-floating celestial object is classified as an unbound planet or a late dwarf star (cf. Table 2.2) is sometimes a matter of appraisal (Liu 2014). In addition to bound and free-floating exoplanets, a population of approximately Jupiter-sized extrasolar planets has recently been described that might not be gravitationally bound to a star (Wambsganss 2011). Given that exoplanets exist, there is no reason why exomoons shouldn't exist likewise. The first possible exomoon, about half the size of Earth's moon, and circling a rogue planet of four Jupiter masses, has in fact been spotted by microlensing (Bennett 2014); for gravitational microlensing, see the following section. To be habitable, a moon and the planet it orbits need not necessarily reside in the habitable zone; as a (giant) planet and its moon(s) are beyond the habitable zone, and the moon is in the "correct" distance from the planet, it can be heated by gravitational (tidal) interaction with the planet to the extent where the moon becomes habitable – in contrast to its cold mother planet. Again, Enceladus is a respective example, as is Jupiter's moon Europa.

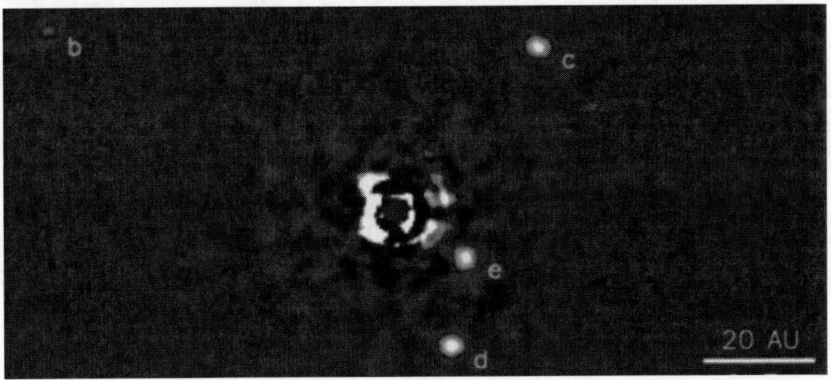

Figure 2.3. Near-infrared image of the four-planet system of the star HR 8799 in the constellation of Pegasus at a distance of 128 ly (Marois 2010); © Nature Publishing Group. The central luminous speckles represent residual light remaining after cutoff of the central star, the tiny circular blots are the four planets. Masses m of the four super-Jupiters (in units of m_J), semi-major axes r (in AU), and orbital period p (in years): HR 8799-b, $m = 7.0$, $r = 68$, $p = 450$; HR 8799-c, $m = 10.0$, $r = 43$, $p = 225$; HR 8799-d, $m = 10.0$, $r = 27$, $p = 112$; HR 8799-e, $m = 9.0$, $r = 14.5$, $p = 49$. The central star is a young ($60 \cdot 10^6$ years) A5 star of 1.56 $m_\odot$. For comparison: Jupiter orbits our Sun at a distance of 5.2 AU in 11.9 years. In addition to the four planets, HR 8799 also possesses a cold and a warm debris belt. HR is short for *Harvard Revised* number, a classification system for bright stars.

2.3. DETECTION METHODS

Most of the ca. 1850 exoplanets confirmed to date have been detected by one of the five detection methods – or combinations thereof – briefly summarized below.

(1) Direct imaging: Hot super-Jupiters emit infrared radiation and can thus sometimes be detected by direct imaging in the (near) infrared, provided that they are in wide orbits, and their bright central star is masked. An example is the system HR 8799-b to HR 8799-e shown in Figure 2.3 in the preceding section. Another example is HD 95086-b (Rameau 2013). Here, the central star is a dusty, early-type A8 star of 1.6 $m_\odot$ in the constellation of Carina, a member of the Lower Centaurus Lux association. The super-Jupiter HD

95086-b (4-5 m_J) orbits its sun at 56 AU; the detection of the planet was carried out at thermal IR (3.8 μm). In many cases, however, planets are outshone by their sun: the sun's brightness commonly exceeds that of the planet by more than a million-fold, rendering direct imaging a severe if not unsolvable problem. Two new detectors recently installed high in the Chilean Andes will focus on young Jupiters and super-Jupiters still glowing from their process of formation. One of the detectors, the Gemini Planet Imager (GPI), is operated by a North-American, the other one, the SPHERE instrument, by a European consortium.

Unbound planets (rouge planets; see the previous section) are also detected by direct methods, provided, of course, that they are sufficiently bright, i.e., retain sufficient heat after having been ejected from their original system. As mentioned above, the limits between a hot Jupiter and a cool dwarf star can be blurred, and a double star system can thus be mistaken for a planetary system.

(2) Transit method: As the planetary system is positioned in such a way that the planet's equatorial orbit lies in the line-of-sight between Earth and the planet's sun, the planet crosses, or transits, the disc of its parent star. Such a transit can be observed as a meticulous periodical drop in the brightness of the star; see Figure 2.4. From the extent to which the star's light dims, the planet's size can be deduced.

In specific cases, information on the planet's atmosphere can also be extracted from spectroscopic analyses of the light passing through the planet's atmosphere. Additional data is available when the planet moves behind the star (so-called "secondary eclipse"): the thermal radiation of the planet disappears as the planet becomes obscured, allowing for an estimate of the planet's temperature. Combined with the radial-velocity method, which allows for the estimate of the mass of the planet, the planet's density and thus details on its physical and geological structure are available.

In a system with more than one planet, perturbation of each planet's orbit gives rise to minute periodical variations in the transit time, a method that is used to confirm a transiting planet, and also to determine the planets' masses (section 4.1).

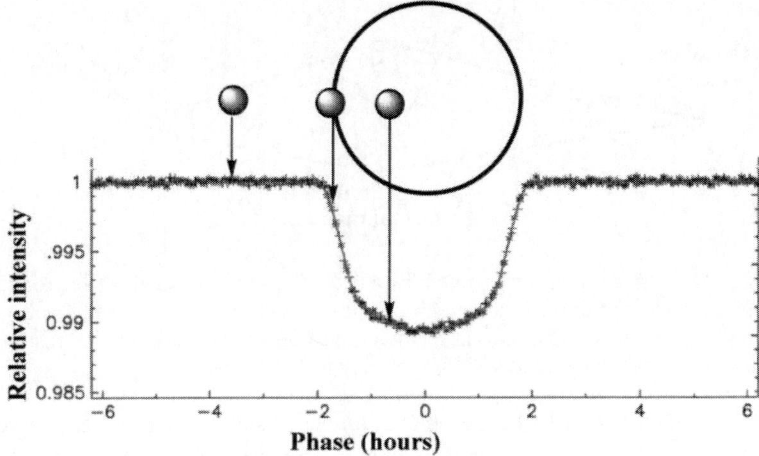

Figure 2.4. Light curve for Kepler 37 (large circle) and its transiting planet Kepler 37-d (small shaded circles; not to scale). Kepler 37 is a star of 0.8 $m_\odot$ (age: 6.0 Gy, surface temperature 5420 K) in the constellation Lyra. The planet's radius is 2 $r_\oplus$, the orbital period 39.8 days (Barclay 2013). Transit light curve: © Reproduced with permission of Nature publishing group.

(3) Radial velocity method: In a binary or multi-component system consisting of a star plus the star's planet(s), the center of gravity is not identical to the star's center of mass. Rather, the planet and its sun move around a common center of gravity, resulting in variations of the star's radial velocity, i.e., the speed by which the star moves back and forth from the Earth in the line-of-sight. This movement results in a Doppler effect (blue shift and red shift, respectively) of the spectral lines of the light emitted by the star. This method of detecting an exoplanet is particularly efficient with low-mass and slowly rotating stars. Stellar activity can deteriorate the reliability of the method.

(4) Gravitational microlensing: When two stars in Earth's line-of-sight are (almost) aligned, the light of the more distant star (the background star) is gravitationally bent and focused by the foreground star, resulting in magnification of the light of the background star. A planet to the foreground star can contribute to this lensing effect when in the appropriate constellation, producing a short spike in the brightness, and thus enabling the detection of this very planet, including its mass and distance from the parent star (see Figure 2.5).

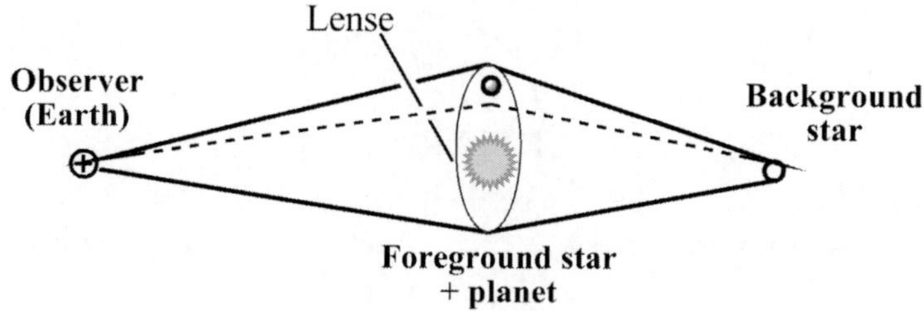

Figure 2.5. Magnification of light emitted by a background star through a foreground star plus its planet. The dashed line represents the situation without a planet. The foreground star (+ planet) functions as a gravitational lense; the respective extension of the light source (the background star) is also known as the Einstein ring.

(5) Carbon monoxide clumps: The extended disks of dust and gas in very young stellar systems can contain extensive amounts of CO. Since the life-time of CO amounts to a few hundred years at the best (CO is liable to UV-induced dissociation), the detection of massive amounts of this gas hints towards an (unseen) planet that replenishes CO from icy debris (grains, comets, planetesimals) in the young sun's (protoplanetary) disk (Dent 2014).

SUMMARY

The stringent definition of a planet in our Solar System, issued by the International Astronomical Union in 2006 (and "degrading" Pluto to a dwarf planet) is not consistently applied to exoplanets: here, the term exoplanets also encompasses unbound (free-floating, or Steppenwolf, planets). Exoplanets can vary in size between sub-Mercuries and super-Jupiters. The upper mass limit for a planet is ca. 13 m_J; beyond that mass, deuterium fusion begins, a property attributed to stars, "brown sub-dwarfs" in this specific case, i.e., stars of low mass and low surface temperature (< 2300 K). Brown dwarfs are subdivided into four categories, viz. M, L, T and Y, with surface temperatures down to 250 K in the case of Y stars. Y stars are at the boundary of exoplanets.

A common scheme for the classification of exoplanets is geared to the planet's size. Accordingly, one distinguishes the following categories: (1) Mercury- to Mars-sized, (2) Earth-sized, (3) super-Earths or mini-Neptunes, (4) Neptune-sized, (5) hot Jupiters, and (6) super-Jupiters. The majority of confirmed exoplanets belong to the categories (3) to (6), and a majority again orbits their star clearly within the orbit of Mercury in our system. This incoherency with our Solar System will possibly be resolved, at least in part, as detection methods improve and theories of the development of planetary systems become refined. Not surprisingly, there is also evidence for the existence of exomoons.

There are four main detection methods for exoplanets: (a) direct imaging, (b) the transit method, (c) variations of the stars' radial velocity, and (d) gravitational micro-lensing. Direct imaging is restricted to large planets at large distances from their sun. The transit method exploits the fact that the brightness of the central star drops when crossed by the planet. The radial velocity method detects the minute wobbling of the star around the system's common center of gravity, resulting in a periodical Doppler shift of the light emitted by the star. Gravitational microlensing is based on the magnification of the light of a background star, aligned with the observer on Earth and a foreground star plus its planet (the "lense").

REFERENCES

Abbot D.S., and E.R. Switzer (2011): The Steppenwolf: a proposal for a habitable planet in interstellar space. *Astrophys. J. Lett.* 735:27-30.

Baily V., T. Mehkat, M. Reiter, et al. (2014): HD 106906 b: A planetary-mass companion outside a massive debris disk. *Astrophys. J. Lett.* 780:L4.

Barclay T., J.F. Rowe, J.J. Lissauer, et al. (2013): A sub-Mercury-sized exoplanet. *Nature* 404:452-453.

Bennett D.P., V. Batista, I.A. Bond, et al. (2014): A sub-Earth-mass moon orbiting a gas giant primary or a high velocity planetary system in the galactic bulge. *Astrophys. J.* 785:155-168.

Cushing M.C., J.D. Kirkpatrick, C.R. Gelino, et al. (2011): The discovery of Y dwarfs using data from the Wide-field Infrared Survey Explorer (WISE). *Astrophys. J.* 743:50.

Dent, W.R.F., M.C. Wyatt, A. Roberge, et al. (2014): Molecular gas clumps from the destruction of icy bodies in the β Pictoris debris disk. *Science* 343:1490-1492.

Dupuy T.J., and A.L Kraus. (2013): Distances, luminosities, and temperatures of the coldest known substellar objects. *Science* 341:1492-1495.

Liu M.C., E.A. Magnier, N.R. Deacon, et al. (2013): The extremely red, young L dwarf PSO J318-22: a free-floating planetary mass-analog to directly imaged young gas-giant planets. *Astrophys. J. Lett.* 777:L20.

Luhman, K.L. (2014): Discovery of a ~250 K brown dwarf at 2 pc from the Sun. *Astrophys. J.* Lett. 786:L18.

Maroi C., B. Zuckerman, Q. M. Konopacky, et al. (2010): Images of a forth planet orbiting HR 8799. *Nature* 468:1080-1083.

Mayor M., and D. Queloz (1995): A Jupiter-mass companion to a solar-type star. *Nature* 378:355-359.

Petigura E.A., A.W. Howard, and G.W. Marcy (2013): Prevalence of Earth-size planets orbiting Sun-like stars. *Proc. Natl. Acad. Sci.* 110:19273-19278.

Rameau J., G. Chauvin, A.-M. Lagrange, et al. (2013): Discovery of a probable 4-5 Jupiter-mass exoplanet to HD 95086 by direct imaging. *Astrophys. J. Lett.* 772:L15.

Tuomi, M. (2012): Evidence for nine planets in the HD 10180 system. *Astronom. Astrophys.* 543:A52.

Wambsganss, J. (2011): Bound and unbound planets abound. *Nature* 473:289-290.

INTERSTELLAR CLOUDS AND PLANETARY SYSTEMS

The space in between the stars that constitutes a galaxy, as well as the intergalactic space, is not empty. Rather, commonly assembled in interstellar nebulae, there is a variety of atoms, molecules and ions, plus (in dense clouds) dust grains that serve as a source for material that forms planetary systems. Dust grains in particular are composed of silica-based and carbon-rich components; large grains are also coated with ice, including water ice, and thus can supply water to the newly formed planetesimals, comets and the developing planets orbiting the young sun. The breakdown of a dense interstellar nebula, triggered for example by a supernova explosion, is commonly the starting point for the formation of a planetary system.

For Life as we know it, water is indispensable. Interstellar water molecules and water ice coating the surface of dust grains are sources for water associated with planetesimals, and the planets that form by accretion of planetesimals in the wake of the formation and development of a planetary system. In the early phase of our home planet, temperatures likely were too high to allow for an appreciable detention of water on Earth's surface. At a later stage, during consolidation of our Solar System, water was replenished to planets in the habitable zone by meteorites (fragments of remainder planetesimals) and comets, stemming from cooler outward regions of the system. The interstellar clouds also have available basic molecules that can serve as building blocks and/or precursors for life molecules; in a way comparable to the supply of water, these molecules can be delivered to a planet orbiting in the habitable zone.

3.1. FORMATION, DEVELOPMENT AND DECAY OF PLANETARY SYSTEMS

Suns and their planetary systems are formed in the wake of collapsing interstellar dark nebulae, also known as molecular clouds. They are termed "dark (molecular) clouds" because they are dark in the visible region; in the far infrared, at wave length > 200 μm, the clouds are pretty luminous. These clouds, or nebulae, comprise several million $m_\odot$ and commonly extend over dozens of light years throughout the galaxies. In our local system, they are visible as irregular patchy patterns obscuring the bright band-like structure at the night sky, the Milky Way, a structure that marks the view towards the edge of our Milky Way Galaxy. An example is the shadowy horsehead nebula in the constellation Orion, embedded in an emission nebula just underneath the left star of Orion's belt (Figure 3.1).

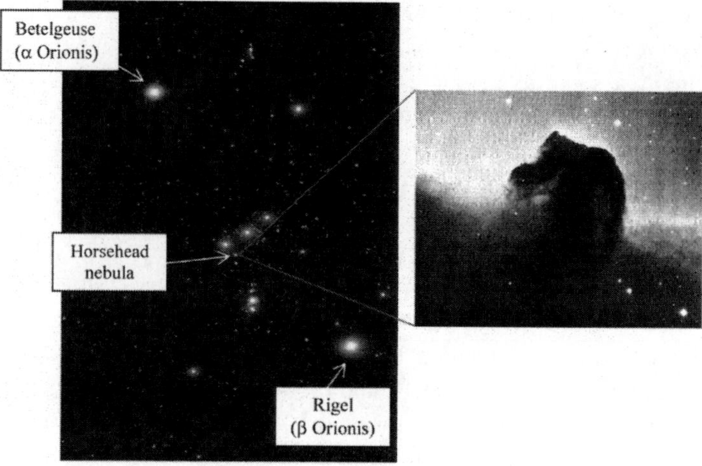

Figure 3.1. The constellation of Orion – a hunter in Greek mythology – with its two most prominent stars indicated (Betelgeuse, the hunter's left shoulder; Rigel, the hunter's right knee) and, enlarged, the horsehead nebula (Barnard[3] 33), just below the left star (Alnitac, or ζ Orionis) of the hunter's belt. Barnard 33 is embedded in the emission nebula *IC 434*.

---

[3] Edward E. Barnard, an astro-photographer, cataloged 370 objects in 1919.

The main constituents of these dark clouds, with temperatures around 10 to 20 K, are hydrogen molecules H_2 and helium atoms He, together amounting to about 99% of the cloud's total mass. Dark clouds also contain a variety of partially complex molecules (section 3.2), some of which are unstable at earthly conditions, and some of which are considered basic molecules for the buildup of life. Of one million atoms, $878 \cdot 10^3$ are hydrogen and $121 \cdot 10^3$ are helium.

The next most abundant elements are oxygen (606 atoms in 1 million), carbon (264), nitrogen (79), magnesium (29), silicon (26), iron (23) and sulfur (13). The overall number density[4] n in these clouds is high (around 10^6 cm^{-3}) when compared to diffuse clouds ($n < 10^3$ cm^{-3}), but very low when compared to Earth's atmosphere ($n \approx 10^{19}$ cm^{-3} at ground level).

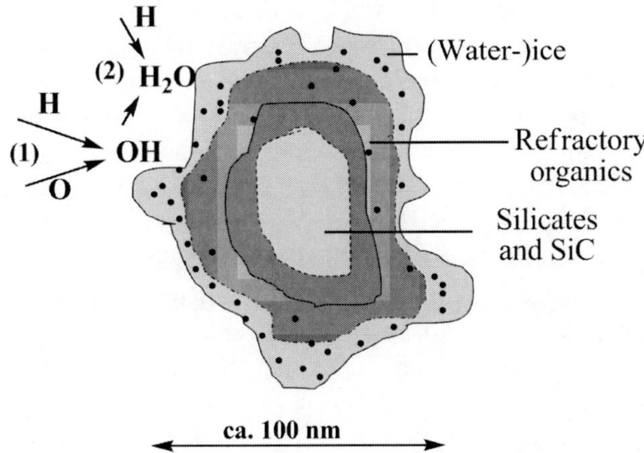

Figure 3.2. The buildup of an average interstellar dust grain. The two shells encapsulating the silicate core, and the black dots, represent "refractory organics", a term that refers to carbon plus carbon-rich organics (such as polyaromates). The shell embracing the silicate core resembles graphite. The grain surface is a site for multiple chemical reactions between atoms and molecules in the cloud. As an example, hydrogen and oxygen atoms can combine to form hydroxyl radicals OH (1); incorporation of an additional H atom (2) results in the generation of water. In both cases, the impact energy and reaction enthalpy are absorbed by the grain, thus preventing the immediate ripping apart of the newly formed molecule.

---

[4] The *number density* $n(X)$, in units of cm^{-3}, specifies the number of atoms, molecules or ions, X, in 1 cm^3. See also "Basic measuring categories and units".

Ca. 1% of the matter of dark clouds is made up of (ice-coated) dust grains, typically ranging in size between 5 and 250 nm. In the context of ice-coated interstellar dust grains, 'ice' refers to a mix of water, carbon dioxide, carbon monoxide, methane, methanol, ammonia, dinitrogen, and other compounds, while the core is composed of silicates, silicium carbides (SiC_x) and refractory, i.e., particularly carbon-rich organics. The general buildup of a typical interstellar dust grain of an approximate size of 100 nm is sketched in Figure 3.2.

Eventually, turbulences and gravitation within the dark clouds cause instability and a breakdown into filamentous substructures, followed by a gravitational collapse, with the gravitational energy being transformed into heat. After a short period of relative stability, the objects further collapse to form a disk-like, rotating pre-planetary nebula. The material aggregates, generating a protosun and a debris belt, accompanied by a dramatic rise in temperature up to 10^7 K, where hydrogen fusion starts in the central protosun. In the wake of further aggregation, meter- to kilometer-sized planetesimals and finally planets emerge (Figure 3.3). Transforming nebular dust to fully formed planets took less than a hundred million years. Our planetary system arose as one of the constituents in a cluster that formed when the interstellar cloud collapsed. Recent evidence strongly suggests that the star HD 162826 in the constellation Hercules is a solar sibling (Ramírez 2014), formed from the same gas and dust cloud. This evidence is based on comparable orbit and movement patterns relative to the galactic center, and on the essentially identical amounts of iron, sodium, aluminum, barium and vanadium in HD 162826 and the Sun.

This so-called core-accretion model for our Solar System is also supported, at least in part, by observations of extrasolar planetary systems. This does not imply that "the Solar System provides a universal template for planetary system architectures" (Howard 2013). Many extrasolar systems have in fact characteristics that are unfamiliar to our system. As an example, planets intermediate in size between Neptune and Earth are very common in extrasolar systems, but absent from our system. And gas giants the size of Jupiter can circle around their suns at distances clearly within Mercury's orbit. Other "extremes" (i.e., planets with characteristics unfamiliar to our planetary system) include WASP 7-b, orbiting its star's poles, or the Kepler 11 system, where comparatively small planets ($m = 2.3\text{-}13.5\ m_\oplus$), orbiting their sun at distances between 0.09 and 0.25 AU, have rather low densities ($\rho = 0.5\text{-}3.1$).

At least several of these "peculiarities" might be explained by planets moving inward towards their sun after formation in more distant regions of their solar system.

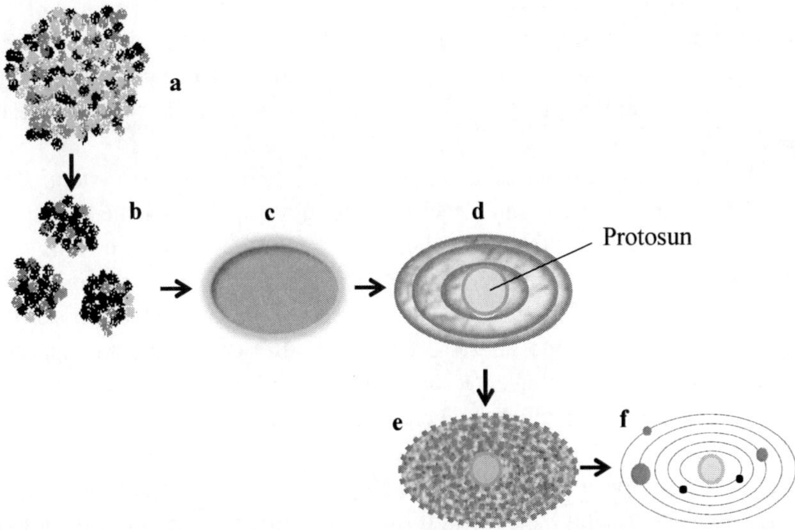

Figure 3.3. The formation of planetary system according to the core-accretion model: An interstellar dark cloud consisting of gas and dust grains (**a**) breaks down into substructures (**b**) that form disk-shaped rotating pre- (or protoplanetary) nebulae (**c**). Electrostatic forces fuse the grains together to form rocky and icy chunks that orbit the protosun (**d**). Eventually, these chunks condense to form km-sized planetesimals (**e**) that further accrete gravity-driven and thus grow into planets (**f**). Friction with the left-over gas and grains forces the planets into orbits of low ellipticity. Close to the protosun, only grains constituted of high melting point materials could survive, resulting in rocky planets with an iron core. Planets that accreted in the outer zones could preserve ices and, when sufficiently massive, also pull in gas from the remains of the nebula, forming giant gas planets such as Saturn and Jupiter in our system.

Planets of about the size of Earth preferentially form at a distance of several AU from the protosun, and can in fact migrate to closer orbits through interaction with the disk. For the formation of Jupiter-sized planets, a sufficiently high density of the protoplanetary disk is afforded: In a first step, a protoplanet forms by accretion of dust and ice. In a successive phase, the protoplanet undergoes runaway gas accretion, provided that the gaseous constituents in the disk have not yet been cleared away. Both the stellar mass and its metallicity correlate with the probability of hosting a giant planet, in that massive stars with high metallicity are likely to host planets larger than Neptune. For the definition of "metallicity" see section 2.1 and footnote 2.

The protostar not only assembles matter, but also ejects matter as a consequence of preserving angular momentum, an essential process to secure the survival of the newly-born star and its constitution as the central sun of the emerging planetary system. Gravitational perturbation by a massive planet, such as Jupiter in our system, can hamper planet formation through its gravitational pull. In that case, the planetesimals in the perturbed region do not accrete.

An illustrative example is the asteroid belt in our Solar System. The planetesimals, or asteroids, can undergo some development in the course of time, including mixing through planetary migration, and are thus not necessarily messengers from the primordial situation. Some of the originary material of the debris belt also passes off to regions far outside the sun and the sun's planets and hence into utterly cold regions. In our system, this material accumulates in the Oort cloud. Occasionally, these chunks become orbitally disturbed and, approaching the Sun, become visible as comets.

An instructive example for the early stage of the formation of a planetary system is the protostar IRAS 04368+2557 (Sakai 2014), embedded in an extended molecular cloud in a star formation region in the constellation of Taurus. This cloud, at a distance of 140 pc, harbors the nearest star formation region. IRAS 04368+2557 is a particularly young protostar, of about $3 \cdot 10^5$ years, with a mass comparable to that of our Sun. The star is surrounded by a flattened, infalling envelope of ca. 1000 AU diameter plus a protostellar disk of ca. 180 AU.

The envelope mainly consists of H_2, but the unsaturated hydrocarbon *cyclo*-propenylidene[5] c-C_3H_2 has also been detected as a comparatively abundant species. In the transition zone between the rotating envelope and the protostellar disk, c-C_3H_2 fades away, possibly by absorption on dust grains. Here, sulfur monoxide SO becomes enhanced.

Along with inorganic molecules, carbon-based organic compounds are delivered by the collapsing cloud and are further processed in the course of protoplanetary disk evolution. Molecules formed in the inner parts of a contracting (protostellar) nebula can flow outward, and be converted, just as in cold dense clouds, into more complex organic and mineral species that become absorbed to ice-coated grains that become eventually incorporated into comets.

---

[5] See sidebar 3.1 for the interstellar formation of c-C_3H_2.

Such a process has been confirmed by the Stardust mission: In the frame of this mission, mineral fragments processed close to the Sun and later integrated into comet Wild 2 have been collected from the coma of the comet and returned to Earth for analysis (Ciesla 2012); see also section 6.1.

During accretion, the main part of the water – originally present as a coating of ice on dust particles (Figure 3.2) – is preserved in the outer cool regions of the accreting system in the form of water ice. In the inner area, water becomes a constituent of the gas phase and thus escapes, leaving the inner planets more or less devoid of water. Also, in the cooler regions of the accreting system, the organic material formed in the clouds and absorbed to the dust particles survives and thus becomes an integral constituent of the new planet. Later, after stabilization of the newly formed planetary system within the first ca. 50 million years, water can be replenished by water-rich planetesimals (coming from planetesimals that did not take part in the accretion process) and by comets, provided that the planet is in a moderate zone, moderate with respect to temperature. These aspects will be dealt with in more detail in section 3.2.

In the early phase of the formation of a planetary system, gravitational interaction of an approximately Earth- or Neptune-sized planet with a gas giant planet can result in the ejection of the less massive planet from the system. Such a rogue, or Steppenwolf, planet drifting unbound through interstellar space could sustain conditions suitable for the development and flourishing of life in subsurface oceans for a few gigayears (Abbot 2011), provided, of course, that the planet formed in a comparatively moderate zone in its native solar system, thus allowing for the retention of the originally incorporated water.

Even after the emergence of the planets and stabilization of the planetary system on the whole, there remains interstellar gas and dust, continuously being captured by the planets. Earth, for example, is constantly bombarded by interplanetary and interstellar dust particles, with an estimated overall daily influx of 60 tons, a significant amount of which enters Earth's atmosphere without being substantially heated, thus enabling the survival of organic matter associated with the dust (Coulson 2006).

The gravitational instability of the genuine interstellar cloud can be triggered by a shock wave, such as that generated by a gigantic star explosion, a supernova. A well-documented example of this type of event is the crab nebula, a supernova remnant, in the constellation of Taurus, going back to a supernova explosion in the year 1054. This explosion, visible even in the day sky, was spectacular to the extent that it became recorded in a wall painting in the Chaco Canyon in the Northwest of New Mexico (Figure 3.4).

While a supernova explosion can initiate the collapse of an interstellar cloud, these titanic explosions also supply massive amounts of dust to the interstellar medium, dust that is rich in amorphous carbon, silicates and iron (Matsuura 2011). But interstellar material is also supplied by outflows of gas – so-called stellar winds – from developed massive stars. Examples are the Wolf-Rayet stars, highly luminous and hot developed stars with masses > 20 $m_\odot$. These stellar winds become increasingly effective as the aging star evolves into a giant star.

Intermediate-mass stars (stars that are insufficiently massive to end up as supernovae) terminate their life by ejecting most of their mass in gaseous outbursts. Part of this gaseous outflow condenses to dust when arriving at cooler regions more distant from the center of eruption (Norris 2012). The *intra*galactic medium is thus continuously enriched with dust grains containing a comparatively large percentage of heavy elements, i.e., elements beyond helium.

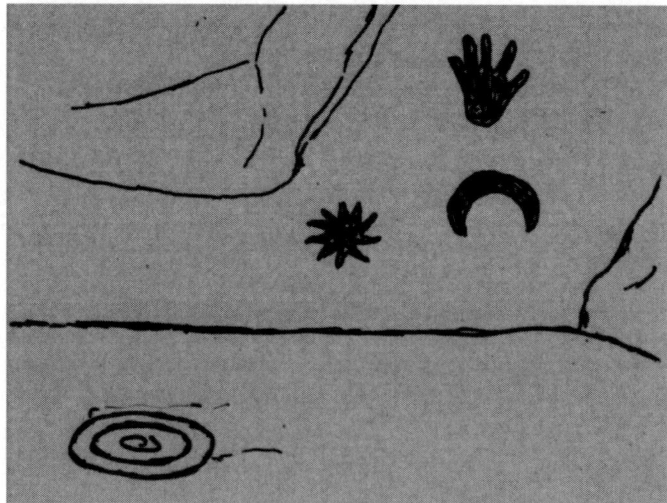

Figure 3.4. The pictograph on a rocky overhang of the Chaco Canyon shows the supernova of July 4, 1054, and the crescent moon – the moon phase at that very day. The hand-print marks a sacred site. The "spiral" on the adjacent wall has been interpreted as Halley's Comet and, alternatively, a symbol for the Sun, the position of which might connect to the summer solstice. Halley's Comet, identified in 1705 by Edmond Halley, is visible from Earth every ca. 76 years.

3.2. FORMATION OF BASIC INTERSTELLAR MOLECULES

In section 3.1, we have briefly dealt with dust grains as a constituent of interstellar clouds, and a main provider for the formation of planetesimals and thus planets. The average size of these dust particles (cf. Figure 3.2) is commonly between 10 and 100 nm. However, the cores of these clouds can also harbor dust grains ten times as large, and hence up to a size of 1 μm on average.

The bulk of the dark clouds' matter is made up of neutral and charged atoms and molecules that also serve as building blocks for dust particles. To date, ca. 180 molecules have been identified in the interstellar galactic medium and in circumstellar shells, 56 of which are also present in the intergalactic space. For websites providing the current status on interstellar, circumstellar and intergalactic molecules, see

www.ph1.uni-koeln.de/node/470
www.cv.nrao.edu/~awootten/allmols.html.

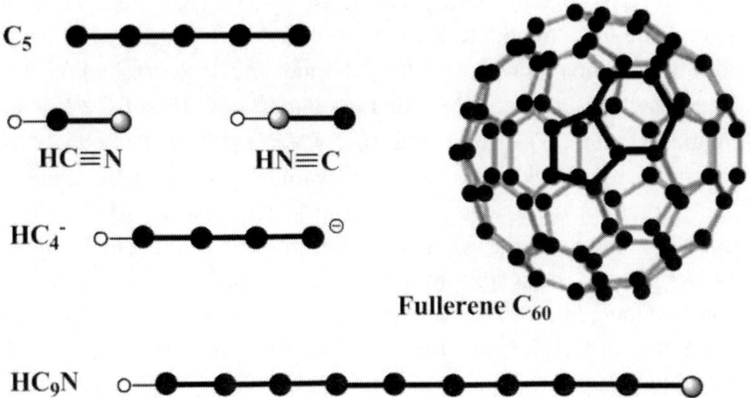

Figure 3.5. A selection of (exotic) molecular carbon-based species in interstellar and circumstellar clouds. One of the indene moieties (5+6-ring) of C$_{60}$ is highlighted with bold connectivities.

Among the molecules and ions detected in the interstellar space (for a selection of carbon-based species see Figure 3.5) there are simple as well as highly complex homonuclear molecular carbon species such as C$_2$, C$_3$, C$_5$, and the fullerenes (also referred to as "bucky balls") C$_{60}$ and C$_{70}$. Eponymous for the spherical molecule C_{60},[6] discovered in a planetary nebula (Cami 2010), was the architect Buckminster Fuller, who became famous in the fifties of the last century for his geodesic dome structures.

In analogy to a soccer ball, C_{60} is composed of 12 convex carbon pentagons and 20 hexagons. Binary interstellar species include the carbide, nitride, sulfide and oxide of silicium (SiC, SiN, SiS and SiO), ferrous oxide (FeO), titanium oxide and dioxide (TiO and TiO_2) sodium chloride (NaCl), and the argon hydride cation $^{36}ArH^+$. The latter was recently detected in the crab nebula (Barlow 2013), a supernova remnant of some interest in two respects: For one thing, molecules of the lighter noble gases, including argon, are non-existent under earthly conditions. For another thing, the argon isotope of mass 36 clearly is indicative of its origin within the frame of a cascade of α captures, starting for example from the carbon isotope ^{12}C. On our planet, by far the more abundant argon isotope is ^{40}Ar (99.6%, making up 0.93% of the dry atmosphere), which formed from the potassium isotope ^{40}K by positron (β^+) emission or electron capture. For α capture, β^+ emission, and electron capture in the context of radioactive processes see sidebar 2.1.

Another "exotic" molecule is hydrogen isocyanide HNC, on Earth just a minor isomer of hydrogen cyanide HCN. In space, HCN and HNC are approximately equally abundant; they are formed *pari passu* via electron capture from the precursor cation $HCNH^+$; cf. sidebar 3.1. Several anions have been detected, such as cyanide CN^-, butadiynylide(1-) C_4H^- ($H-C\equiv C-C\equiv C^-$), and cyanobutadiynylide(1-) NC_5^- ($N\equiv C-C\equiv C-C\equiv C^-$). Interstellar clouds also harbor several neutral cyanides and isocyanides of a more complex composition, including the linear eleven atom species HC_9N ($H-C\equiv C-C\equiv C-C\equiv C-C\equiv C-C\equiv N$). Of particular interest in the context of key molecules as precursors for more complex molecules essential for life, such as phospholipids, sugars, nucleobases and peptides, are the species methylsulfide (methylmercaptane) CH_3SH, methanol CH_3OH, formaldehyde HCHO, hydrogen cyanide HCN, and aminoacetonitrile H_2NCH_2CN. We will discuss this issue in some detail in section 6.1.

Additional interstellar matter, though of minor interest with respect to molecules essential for life, is present in ionized and neutral diffuse clouds with number densities $n(X)$ of around 10^2 to 10^3 cm^{-3} and temperatures of 10^4

---

[6] The fullerenes were first generated in 1985 by a team of chemists (Kroto et al.; Nobel prize 1996) through laser evaporation of graphite, in search of preparative routes to interstellar oligoacetylenes.

K (ionized clouds; HII regions) or 50-100 K (neutral diffuse clouds; HI regions). Furthermore, a considerable amount of overall matter is present *beyond* the galaxies. A large region around the Milky Way, extending to more than 100 kpc, harbors some 10^{10} $m_\odot$ of highly charged (and hence "hot") ions, such as OVII (= O^{6+}) and OVIII (= O^{7+})[7], although with very low column densities $N(X)$[8] of ca. 16 cm^{-2} (Gupta 2012). This hot intergalactic matter, also termed intergalactic coronal gas, is of quite some interest in the wake of chemistry at particularly extreme conditions in outer space, but appears to be less meaningful when it comes to the creation of molecules relevant for life.

In contrast, rich chemistry with a potentiality for life takes place in the comparatively dense dark molecular clouds, where number densities $n(X)$ are up to 10^6 cm^{-3}, and temperatures are below 20 K. Compared to the average conditions in our atmosphere – a number density of ca. $3 \cdot 10^{19}$ cm^{-3} at sea level and a temperature around 290 K – this is still close to vacuum and absolute zero. In a dark cloud, the mean free path of an atom or molecule, and thus the chance to meet a potential reaction partner, ranges from several to many dozens of years.

And even as two potential reaction partners approach each other sufficiently close to 'recognize' their mutual presence and thus to undergo a chemical reaction and to form a (new) molecule, the resulting molecule is commonly instantaneously torn apart, since there is no way to dissipate the impact energy. The situation for an effectual reaction, where the impact energy is carried away by a second reaction product, is illustrated symbolically in Figure 3.6.

One way to circumvent this problem is a third partner in the collision who absorbs the energy (option 1), or the generation of a second product of reaction that carries off the impact energy. This second "reaction product" can be an atom/ion/molecule (option 2a), an electron (option 2b), or electromagnetic radiation hν (option 2c).

---

[7] In astronomical issues, ions are commonly indicated by the element symbol plus a Roman number: I for atoms (neutral), II for ions carrying one positive charge, III for ions with two positive charges, etc. Examples are HI ≡ H, HII ≡ H$^+$, OIII ≡ O^{2+}. Interstellar HI regions are thus areas dominated by hydrogen atoms, HII regions those where hydrogen ions prevail.

[8] The *column density* $N(X)$, in units of cm^{-2}, indicates the number of atoms etc. in a column of 1 cm^2 cross section. "Column" refers to the length of the line-of-sight between an observer and a light-emitting object, such as a star or a distant galaxy [in atmospheric chemistry: the height of the atmosphere]. For the distinction from *number density* $n(X)$, see footnote 4 and "Basic measuring categories and units".

Option 1 has been illustrated in Figure 3.2: here, the impact energy becomes dissipated via absorption by a dust grain. Another prominent example is the formation of molecular hydrogen H_2, Eq. (3.1). Examples for option (2) are provided in Eqs. (3.2)-(3.4).

In Eq. (3.2) an oxygen ion and a hydrogen molecule combine to form a hydroxy cation and a hydrogen atom. Eq. (3.3) represents the formation of the two isomers of hydrogen cyanide, and Eq. (3.4) the interstellar radiation-driven generation of a neutral hydrocarbon and hence a member of a comparatively large family of "organic" interstellar compounds

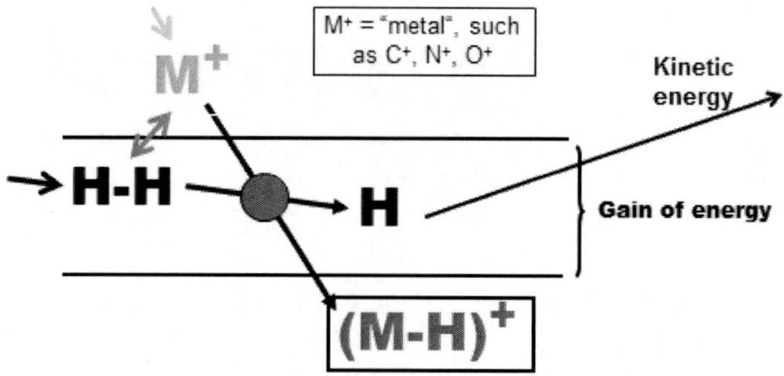

Figure 3.6. The formation of a MH$^+$ species in interstellar clouds. The cation M$^+$ and the molecule H_2 'sense' each other (double arrow) in a close encounter: M$^+$ induces a dipole moment in H_2, resulting in electrostatic attraction and finally collision of the two reaction partners (dark-grey circle). The intermittently formed $\{H_2M\}^+$ (not shown) instantaneously decays to form the new molecular cation (MH)$^+$ and hydrogen atoms H. The net reaction, $M^+ + H_2 \rightarrow MH^+ + H$, is exergonic and thus thermodynamically driven; energy gain and impact energy are carried off by the hydrogen atom H in the form of kinetic energy (energy of motion).

For a more systematic overview of reaction types in interstellar clouds, including examples, see sidebar 3.1.

$$H + H + \text{grain} \rightarrow H_2 + \text{grain} \tag{3.1}$$

$$O^+ + H_2 \rightarrow OH^+ + H \tag{3.2}$$

$$H_2NC^+ + e^- \rightarrow (\text{HCN and HNC}) + H \tag{3.3}$$

$$C_2H^- + h\nu \rightarrow C_2H + e^- \tag{3.4}$$

The low temperature in interstellar clouds also presupposes that chemical reactions are kinetically rather than thermodynamically controlled, and further that endothermic reactions (reactions in which energy is consumed) are out of range. Chemistry in dark clouds is often initiated by cosmic rays. Cosmic rays are highly energetic, positively charged particles that easily penetrate – contrasting electromagnetic radiation – dark clouds. Cosmic rays mainly consist of protons and α particles (helium nuclei), plus a small fraction (ca. 1%) of heavier, highly-charged cations such as those derived from carbon and iron.

***

Sidebar 3.1. Reaction types in interstellar clouds

Along with the reactions on grain surfaces (Figure 3.2 and Eq. (3.1) in the main text), there are various types of reactions, involving atoms, molecules, cations and anions, electrons and electromagnetic radiation ($h\nu$). Owing to the abundance of hydrogen, the most common reaction partners are the hydrogen species H, H^+, H^-, H_2, and H_3^+. Hydride H^- forms by electron attachment according to $H + e^- \rightarrow H^- + h\nu$.

Ionization of hydrogen to form H^+ (and helium to form He^+) is effected by cosmic rays. The noble gas helium does not form long-lasting molecules. Next to hydrogen, the most prominent elements are oxygen, carbon and nitrogen. These elements with comparatively low ionization potential can be ionized by the UV of nearby stars, or new-born stars embedded in the clouds. They are therefore present mainly in mono-cationic form, i.e., O^+, C^+ and N^+.

In the following compilation, examples are provided for reaction types that are characteristic for interstellar clouds (Rehder 2010; Wakelam 2010).

Key species from reactions leading into the families of interstellar molecules based on carbon, nitrogen and oxygen are highlighted in bold.

Ion-neutral reactions: $H_3^+ + N \rightarrow NH_2^+ + H$; $H_3^+ + O \rightarrow OH^+ + H_2$; $C^- + NO \rightarrow CN^- + O$

Anion-cation recombination: $HCO^+ + H^- \rightarrow H_2 + CO$

Neutral exchange: $CH + O \rightarrow CO + H$

Dissociative recombination: $H_2CO^+ + e^- \rightarrow CO + 2H$

Dissociative attachment: $CH + O \rightarrow HCO^+ + e^-$

Associative detachment: $C^- + H_2 \rightarrow CH_2 + e^-$

Radiative recombination: $H_2CO^+ + e^- \rightarrow H_2CO + h\nu$

Radiative association: $C_4H + e^- \rightarrow C_4H^- + h\nu$ (electron capture);

$C^+ + H_2 \rightarrow CH_2^+ + h\nu$ (ion-molecule reaction); $C + H_2 \rightarrow CH_2 + h\nu$ (neutral)

Photodissociation: $CO + h\nu \rightarrow C + O$

Photodetachment: $C_2H^- + h\nu \rightarrow CH_2 + e^-$

Charge transfer (resonance charge exchange): $P^+ + H_2S \rightarrow P + H_2S^+$

There are numerous interconnections within the manifold of reactions. Potential reaction sequences providing neutral interstellar molecules that are pivotal in the generation of life molecules (formaldehyde HCHO, methylsulfide CH_3SH, and hydrogencyanide HCN) are shown below.

$$C^+ + H_2 \longrightarrow CH^+ + H$$
$$\xrightarrow{H_2} CH_2^+ + H \xrightarrow{H_2} CH_3^+ + H \xrightarrow{S} H_2CS^+ + H \xrightarrow{H_2} CH_3S^+ + H \xrightarrow{H_2} CH_3SH^+ + H \xrightarrow{e^-} \boxed{CH_3SH}$$
$$\xrightarrow{e^-} CH_2 + H \xrightarrow{OH} \boxed{HCHO}$$

$$N^+ + H_2 \longrightarrow NH^+ + H$$
$$\xrightarrow{H_2} NH_2^+ + H \xrightarrow{H_2} NH_3^+ + H \xrightarrow{Mg} NH_3 + Mg^+ \xrightarrow{C^+} HCNH^+ + H \xrightarrow{e^-} \boxed{HC\equiv N + C\equiv NH} + H$$

For the formation of *cyclo*-propenylidene $c\text{-}C_3H_2$ in, e.g., the envelopes of protoplanetary nebulae (section 3.1), see the following detail of the reaction sequence; *l* and *c* indicate *l*inear and *c*yclic, respectively:

$$CH_2 + C^+ \longrightarrow C_2H^+ + H$$

$$\downarrow H_2$$
$$C_3H^+ + H$$

$$\downarrow H_2$$
$$l\text{-}C_3H_3{}^+ \rightleftharpoons c\text{-}C_3H_3{}^+ + h\nu$$

$$\downarrow e^-$$
$$\boxed{c\text{-}C_3H_2} + H$$

$$c\text{-}C_3H_2 \; = \; \underset{C}{HC{=}CH}$$

***

SUMMARY

Interstellar (*intra*galactic) dark molecular clouds, or nebulae, are composed of about 99% molecular hydrogen (H_2) + helium atoms (He), and ca. 1% differentiated dust grains. Some of this matter is emitted into space from circumstellar clouds ejected by evolving and by dying stars. The maximum particle density amounts to about 10^6 per cubic centimetre. Roughly 180 molecules and ions have been detected, including molecules that can serve as building blocks and/or precursors for more complex molecules that constitute the matter of life. Approximately 56 molecular species are present in the extremely dilute *inter*galactic space.

The low temperature of just 10-20 K in the dark clouds excludes endothermic reactions. Furthermore, a third reaction partner is afforded to carry off the impact energy and the energy of reaction. Other atoms/molecules, dust grains, and also electrons and electromagnetic radiation (hν) can come in as reaction partners. Dust grains, typically ranging in size between 10 and 100 nm, are of particular interest in this context. Larger grains are differentiated in that they contain a core of silicates plus silicium carbide, surrounded by carbonaceous matter which in turn is coated by ice, including water ice.

The interstellar dark clouds are the birth place of planetary systems: The clouds are gravitationally unstable and collapse, triggered for example by a supernova explosion. As a result, the cloud's dust and molecular constituents accumulate within a spiralling disc plus a central protosun sufficiently hot to enable nuclear fusion. In order to preserve momentum, the protosun ejects part of the matter back into the disk. Within the disc, matter gravitationally accretes to form planetesimals, rocky and icy chunks with dimensions up to several kilometers. From these planetesimals, planets accrete. In the course of roughly 50 million years, the planetary system runs through a first phase of consolidation and thus stability.

As, in a later phase of the system, a planet orbits its sun at a distance where water can be present in the liquid state of aggregation, the planet is referred to as being in the "habitable zone". Earth, and very likely also Mars, is an example of this situation in the Solar System. Water is a comparably abundant molecular species in the dark clouds, and thus also in the planetesimals, planets and other bodies, comets in particular. Other interstellar molecules, essential for the buildup of complex molecular systems that constitute Life, include formaldehyde H_2CO, hydrogen cyanide HCN, and methyl sulfide CH_3SH.

REFERENCES

Abbot D.S., and E R. Switzer (2011): The Steppenwolf: a proposal for a habitable planet in interstellar space. *Astrophys. J. Lett.* 735:27-30.

Barlow M.J., B.M. Swinyard, P.J. Owen, et al. (2013): Detection of a noble gas molecular ion, $^{36}ArH^+$, in the crab nebula. *Science* 342:1343-1345.

Cami, J., J. Bernard-Salas, E. Peeters, and S.E. Malek (2010): Detection of C_{60} and C_{70} in a young planetary nebula. *Science* 659:1180-1182.

Ciesla, F.J., and S.A. Sanford (2012): Organic synthesis via irradiation and warming of ice grains in the solar nebula. *Nature* 336:452-454.

Coulson S.G. (2006): Meteorides: a landing capsule for panspermia. *Int. J. Astrobiol.* 5:307-312.

Gupta, A., S. Mathur, Y. Krongold, et al. (2012): A huge reservoir of ionized gas around the Milky Way: accounting for the missing mass? *Astrophys. J. Lett.* 756:L8.

Howard A.W. (2013): Observed properties of extrasolar planets. *Science* 340:572-576.

Matsuura M., E. Dwek, M. Meixner, et al. (2011): Herschel detects a massive dust reservoir in supernova 1987A. *Science* 333:1258-1261.

Norris, B.R.M., P.G. Tuthill, M.J. Ireland, et al. (2012): A close halo of large transparent grains around extreme red giant stars. *Nature* 484:220-222.

Pagani L., J. Steinacker, A. Bacmann, et al. (2010): The ubiquity of micrometer-sized dust grains in the dense interstellar medium. *Science* 329:1622-1624.

Ramírez, I., A.T. Bajkova, V.V. Bobylev, et al. (2014): Elemental abundances of solar sibling candidates. *Astrophys. J.* arXiv:1405.1723vl

Rehder, D. (2010): Chemistry in Space – From interstellar matter to the origin of Life. Wiley-VCH, Weinheim.

Sakai, N., T. Sakai, T. Hirota, et al. (2014): Change in the chemical composition of infalling gas forming a disk around a protostar. *Nature* 507:78-80.

Wakelam, V., I.W.M. Smith, E. Herbst, et al. (2010): Reaction networks for interstellar chemical modelling: improvements and challenges. *Space Sci. Rev.* 156: 13-72.

Chapter 4

ORGANIZATION AND TEXTURE
OF EXOPLANETS

About 15% of sun-like stars (G- and K-type; cf. sidebar 2.2) host planets with 1-3 $m_\oplus$, commonly orbiting their sun on low eccentricity and low inclination orbits within 0.25 AU, and thus inside Mercury's orbit (0.39 AU) in our system, i.e., clearly beyond the habitable zone. Densities of approximately Earth-size objects, of super-Earths (up to 2 Earths) and mini- or sub-Neptunes (2-4 Earths)[9] can vary considerably, namely from 9 g cm^{-3} (an example is Kepler 10-b) to 0.5 g cm^{-3} (Kepler 11-e). The rather high mean density of Kepler 10-b suggests that the planet's setup is exclusively from iron and rock, and that it is devoid of an appreciable atmosphere. By contrast, the low density of planet Kepler 11-e is suggestive of an extended atmosphere formed by low molecular mass gases. For the composition of planets with intermediate densities there are several options, namely variants including (i) a rock/iron core surrounded by a light-gas atmosphere; (ii) a rock/iron core immersed in an extended water ocean; and (iii) a mélange of these three components (i.e., rock, iron and water), with an atmosphere dominated by H_2 and He (Howard 2013a). A planet orbiting its sun in the habitable zone – where "habitable zone" again is defined as that distance from the sun where liquid water can exist – this planet is of particular interest with respect to the possible existence of life as we know it. The phrase "Life as we know it" excludes the feasibility of alternative life forms *independent* of water, and also

---

[9] For a more detailed discussion devoted to the distinction between super-Earths and sub-Neptunes with respect to the planets' internal structure and the composition of their atmosphere, see E. Jawin: Super-Earths versus Sub-Neptunes Part 2, Interior Structure: Exoplanets, 2014.

does not take into account the *probability* for life to develop at all. These two aspects will be approached in some detail in Chapter 5.

Apart from the Earth- to approximately Neptune-sized objects that, to some extent, resemble in composition and buildup the respective planets in our system, there are Jupiter-sized planets and super-Jupiters with up to 13 m_J, and a composition probably not unlike the genuine Jupiter in our system. But there are also "hot Jupiters", i.e., about Jupiter-sized planets in orbits extremely close to their sun, with a composition and intra-planetary dynamics that, in all probability, greatly differ from what we know about Jupiter in the Solar System.

In the following sections, we will elucidate the recently emerging knowledge of the architecture of exoplanets, including their atmospheres. The overall information is still scarce, owing to limitations in instrumentation when it comes to distant objects as tiny as planets. Many conclusions, drawn for example on the basis of the global density of an exoplanet, are of a provisional nature since they are based on indirect evidence, and commonly guided by and extrapolated from what is known of our home system. In the context of planetary composition, we will also briefly refer to the composition and properties of extraterrestrial material falling to Earth, namely meteorites.

4.1. GEOLOGICAL AND PHYSICAL FEATURES

A crucial point in revealing a planet's physical and chemical buildup and structure is the knowledge of its size *and* mass. In section 2.3, we briefly summarized the principal methods that can be applied in detecting exoplanets. These methods usually reveal the planet's size and orbital period. For large objects such as planets of the size of Jupiter and beyond, masses can be estimated by measuring the wobble of a star as it moves relative to the system's (star plus planet) common center of gravity. An alternative method providing a clue to the planet's mass, also applicable with some accuracy to Earth-sized planets, is "transit timing variation": As two or more planets orbit a star, their mutual gravitational tugs give rise to minor alterations in their orbital periods. When the two planets *transit* their star and are thus detectable via the periodical dimming of the star's light, the planetary masses can be estimated from the time rate of changes in the intensity of the star's light.

Examples are the planets KOI 314-c and KOI 314-b (Kipping 2014). KOI 314 is a dim dwarf (an M1 type star of 0.57 $m_\odot$). The radii of both planets are ca. 1.6 $r_\oplus$; they are thus somewhat larger than Earth. The orbital periods are 23 and 14 days, respectively, i.e., there is a 5:3 orbital resonance between the two objects. KOI 314-c has approximately the mass of Earth; its sibling KOI 314-b is four times more massive. The surface temperature of KOI 314-c is 104 °C. This is at the limit for potential life as we know it; on Earth, extremophiles can tolerate up to ca. 120 °C. The composition of KOI 314-c does not have much in common with Earth: its average density is 1.3 ± 0.7 g cm^{-3}, and thus only slightly more than the density of liquid water, hinting towards a rocky core and a particularly extended gaseous envelope, taking about 17% of the planet's radius.

A second method to determine the mass of an approximately Earth-sized planet is based on ultra-high-precision Doppler spectroscopy (Howard 2013b), i.e., the red-shift and blue-shift of the light emitted by the parent sun in response to the sun's wobbling caused by the planet's orbital motion. An example of a planet discovered by this method is Kepler 78-b. This planet orbits its parent sun Kepler 78 (a comparatively young star of 0.83 $m_\odot$ and an effective temperature of 5140 K in the constellation Cygnus) at a very low distance – less than two stellar radii – with an orbital period of only 0.355 days. From its mass (1.69 $m_\oplus$) and radius (1.17 $r_\oplus$) a density of $\rho = 5.57$ g cm^{-3} derives, hence a mean density comparable to that of Earth ($\rho = 5.515$ g cm^{-3}), suggesting a similar composition, namely an iron core and a rocky mantle. However, the high surface temperature of around 2300 K precludes a low molecular mass atmosphere and water.

Some of the recently discovered multi-planet systems also harbor Earth-sized planets. Representatives are the systems Kepler 20 (Fressin 2012) and Kepler 62 (Borucki 2013). For both systems, in the constellation Lyra, five planets have been identified, two of which have radii comparable to our home planet, and bulk compositions that could be similar to Earth's. Some of the planets' characteristics are listed in Table 4.1. Kepler 20 is a G8 spectral type star and thus is roughly comparable to our Sun (G2, $T = 5780$ K), while the K2 star Kepler 62 is somewhat smaller ($r = 0.63$ $r_\odot$) and cooler ($T = 4925$ K) than our Sun. The planets Kepler 20-e and Kepler 20-f orbit their central star closely; they are rocky, with an estimated composition of about 32% iron and 68% silicate. The super-Earths Kepler 62-e and Kepler 62-f revolve around their sun more distantly. Both are dry and rocky or, alternatively, consist of essentially frozen water surrounding a silicate and iron core. With $7 \cdot 10^9$ years, the Kepler 62 system is approximately twice as old as the Solar System ($4.6 \cdot 10^9$ years), where Life is believed to have started ca. 3.8 billion years ago. The solar flux received by the planets Kepler 62-e and Kepler 62-f is 1.2 and 0.4 times, respectively, the solar flux for Earth. These two planets are thus in the habitable zone (see Figure 4.1). Similar considerations apply to the recently confirmed planet Kepler 186-f (Quintane 2014).

Table 4.1. Selected properties of Earth-sized exoplanets

	Orbital period (days)	Mean orbital radius (AU)	Planetary radius (in units of $r_\oplus$)	Maximum mass (in units of $m_\oplus$)	(Surface) temperature, (K)
Kepler 20-e	6.1	0.051	0.87	1.67	1040
Kepler 20-f	19.6	0.11	1.03	3.04	705
Kepler 62-e	122.4	0.427	1.61	36	270±15
Kepler 62-f	267.3	0.718	1.41	35	208±11
Kepler 186-f	129.9	0.37	1.1	1.44[a]	~273

[a]For an Earth-like composition (1/3 iron, 2/3 silicate).

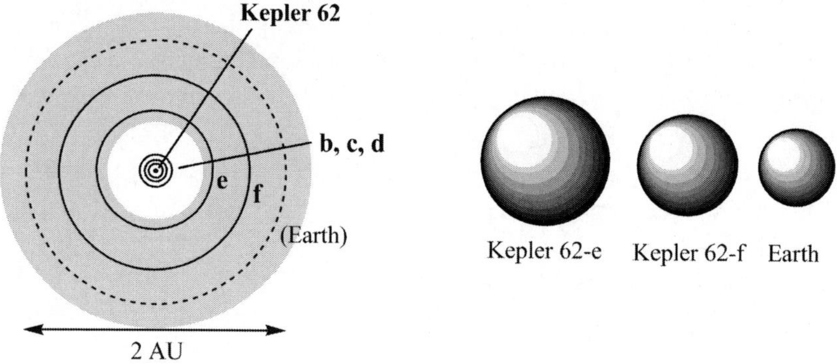

Figure 4.1. The system Kepler 62 with its five planets (b, c, d, e, and f). The super-Earths Kepler 62-e and 62-f are in the habitable zone (grey). For comparison, Earth's orbit around the Sun has been included as a dashed circle. The relative sizes are shown to the right. The radius of the orbit of Kepler 62-e (0.427 AU) is close to the semi-major axis of Mercury (0.387 AU) in our system. While Mercury is not in the habitable zone, the habitable zone for Kepler 62 extends to the orbit of Kepler 62-e because the star is distinctly cooler (the surface temperature is 4925 K) than our Sun (5780 K).

4.2. CHEMICAL COMPOSITION AND ATMOSPHERES

Generally speaking, surveys of extrasolar planets provide approximate information on their size (radius), as well as orbital characteristics such as the semi-major axis, eccentricity, and the orbital period. Masses and consequently, in connection with the size, mean densities are less reliably available, limiting conclusions with respect to the bulk composition. Still less accessible is the specification with respect to atmospheric and surface composition, and hence information that allows for judging whether a planet is habitable. Some indication into this direction is, however, obtained by extrapolating from the information we have available on our Solar System. As for the composition of atmospheres, this information is assembled in Table 4.2.

Table 4.2. Main gaseous constituents of the atmospheres of the planets in the Solar System, Jupiter's moon Io, and Saturn's moon Titan

	Pressure (Pa) at ground level	Main atmospheric constituents (%)
Mercury	10^{-9}	No appreciable atmosphere
Venus	$9.2 \cdot 10^{9}$	CO_2 (96.5), N_2 (3.5), SO_2 (0.015)
Earth	10^{5}	N_2 (78.1), O_2 (20.1), H_2O (around 1), Ar (0.9), CO_2 (0.04)
Mars	$0.6 \cdot 10^{3}$	CO_2 (95.7), N_2 (2.7), Ar (1.6), O_2 (0.13), CO (0.07), H_2O (0.03)
Jupiter	Not defined	H_2 (90), He (9), CH_4 (0.3), H_2O (0.1), NH_3 (0.026)
Io	10^{-4}	SO_2 (90), SO, NaCl, S, O_2
Saturn	Not defined	H_2 (94), He (5), CH_4 (0.2), H_2O (0.1), NH_3 (0.01)
Titan	$1.47 \cdot 10^{5}$	N_2 (98.4), CH_4 (1.4), H_2 (0.15)
Uranus	Not defined	H_2 (83), He (15), CH_4 (2.3), NH_3 (0.01)
Neptune	Not defined	H_2 (82), He (17), CH_4 (1.5), HD (0.02), NH_3 (0.01)

There is no compelling reason why concepts such as habitability should grossly deviate from what we experience in our home system. Nonetheless, the planetary systems so far discovered, including the multi-planetary systems, appear to be rather different in overall arrangement from our system, in many (but not all) cases including the nature and the age of the central sun. Information on extrasolar planets will change as time evolves, and instrumentation for detecting planets plus methods for evaluating the data sets improve, allowing for the detection of, for example, Earth-sized planets in orbits where temperature and radiation conditions are comparable to those immanent for Earth.

The metallicity (see section 2.1/footnote 2 for a definition) of the host star is a proxy for the initial solid inventory of the protoplanetary disc, and is thus correlated to the setup of the planet (Buchhave 2014). Exoplanets have thus been categorized into three populations: (i) Metallicity -0.02±0.02: earth-like planets with a radius $r > 1.7\ r_{\oplus}$; (ii) metallicity +0.05±0.01: gas-dwarf planets with rocky cores and a hydrogen-helium envelope, $r = 1.7$–$3.9\ r_{\oplus}$; and (iii) metallicity +1.08±0.02: ice or gas giant planets, $r > 3.9\ r_{\oplus}$.

As noted in the preceeding section, exoplanets ranging in size from Mercury to super-Earths/mini-Neptunes commonly have internal structures and densities matchable to the inner planets in our system. Roughly, this means that there is a (partially liquid) iron-dominated core, and a rocky mantle and crust, the latter also harboring appreciable amounts of water of crystallization, liquid water and/or water ice (Figure 4.2). "Water of crystallization" refers to "hidden" water contents in minerals, i.e., water incorporated into the crystal lattice of a mineral. Examples are gypsum ($CaSO_4 \cdot 2H_2O$), H_2O/OH^- interlayers in clays, interstitial water in zeolites, or ringwoodite; for concealed water in ringwoodite.

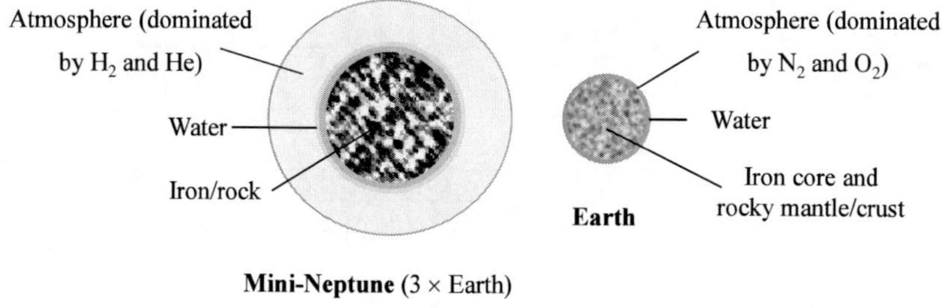

Figure 4.2. Comparison of the size and composition of a mini-Neptune with Earth.

As for Earth, Mercury and the early Mars, the outer layer of the iron core of an Earth-like exoplanet supposedly is liquid. Such a rotating electrically conducting fluid in between two solid layers (the solid inner iron core and the rocky mantle) generates – based on the Dynamo effect – a permanent magnetic field, sufficiently fast rotation of the planet provided. Venus is an example of a planet with a buildup corresponding to that of Earth; however, it has *slow* rotation and is thus *devoid* of a permanent magnetic field.

In contrast, gas giants such as Saturn and Jupiter, when at distances from their sun clearly beyond the habitable zone (as in our system), accrete in the protosolar system from planetesimals assembled from dust as well as methane and noble gases trapped in the form of icy water clathrates[10], providing roughly 5-15% of the overall mass. This accretion is followed by gravitational

---

[10] Clathrates are host-guest compounds, here cages of water molecules encapsulating methane. In a typical methane hydrate, the oxygens of twenty water molecules are connected to a pherical pentagon-dodecahedron encasing one methane molecule. For details see Figure 5.9 in section 5.3.

trapping of additional gaseous constituents from the original nebula. Heavier elements are enriched in these planets relative to the heavy metal contents of the central sun, while helium is depleted in the planets' atmospheres due to condensation onto metallic hydrogen, overlaying the dense rocky core: At pressures > 300 GPa, hydrogen is in a degenerate state with "freely floating" electrons providing conductivity comparable to metallic conductance.

As in the case of Jupiter and Saturn (Table 4.2), the atmospheres of these giant planets are dominated by hydrogen (about 90%) and helium (around 7%); in addition, there is some H_2O CH_4 and NH_3. This atmospheric composition, plus some CO, is also characteristic of T-type brown dwarfs. By way of example, the brown dwarf double system Luhman 16AB consists of two objects of the size of Jupiter: Luhman 16B shows dark areas due to thick clouds (Crossfield 2014). Minor atmospheric components are water, methane CH_4 and ammonia NH_3, plus some hydrogen sulfide H_2S, phosphane PH_3, arsane AsH_3, and germane GeH_4. By reaction of these hydrides with water, the oxides arsenic As_2O_3 and germanium dioxide GeO_2 can form. The reaction between germane and H_2S provides germanium sulfide GeS. These oxides and sulfides give rise to the formation of inorganic particulate matter suspended in the atmosphere. Organic particulate matter based on (poly)hydrocarbons can be produced by (photo-)reactions starting from methane.

These scenarios outlined for approximately Jupiter-sized planets residing clearly beyond the habitable zone only partially – if at all – apply to super-Jupiters, hot Neptunes and hot Jupiters. Super-Jupiters have characteristics approaching those of Y-type brown dwarfs (see Table 2.2). Hot Neptunes and hot Jupiters are planets in orbits very close to their sun, commonly clearly inside Mercury's orbit, namely at distances of 0.01 to 0.1 AU. Here, chemical species in the atmospheres are highly affected by the central star's UV flux. The main constituents in the atmosphere of these hot planets are H, H_2, O, H_2O and CO_2; for somewhat cooler planets, CH_4 becomes dominant. CH_4 and H_2O are particularly sensitive to UV irradiation, i.e., these species disappear in the outer parts of the atmosphere. This is demonstrated for the dissociation of water by UV light (symbolized by hν) in Eq. (4.1a). Successive reactions lead to a depletion of the primarily formed hydroxyl OH, Eq. (4.1b), and a limited formation of carbon dioxide, Eq. (4.1c) (Miguel 2014).

$$H_2O + h\nu \rightarrow H + OH \tag{4.1a}$$

$$H + OH \rightarrow O + H_2 \tag{4.1b}$$

$$OH + CO \rightarrow CO_2 + H \tag{4.1c}$$

In analogy to the supply of water to the inner planets (Mercury, Venus, Earth, Mars) and to dwarf planets (Eris, Vesta) in our system, the main part – if not all – of the water present on exoplanets in moderate (including habitable) zones originates from comets and meteorites. In addition, chemical sputtering comes in. In such a process, oxygen atoms O and ions O^+ are blasted off the surface of minerals, signified in Eq. (4.2) by {MO} (M for metal, e.g., Fe, Al and Mg), by highly energetic protons mainly stemming from the solar wind. The extremely thin exosphere of Mercury is of comparable origin. Once O^+ is generated, a reaction cascade as represented by Eq. (4.3) goes on, reminiscent of reactions in interstellar clouds (see sidebar 3.1).

$$\{MO\} + H^+ \longrightarrow O^+ + \{M\} + H \tag{4.2}$$

$$
\begin{aligned}
O^+ + H_2 &\longrightarrow OH^+ + H \\
&\quad\Big\downarrow{}^{H_2} \\
&\qquad\quad H_2O^+ + H \\
&\qquad\quad\Big\downarrow{}^{H_2} \\
&\qquad\qquad\quad H_3O^+ + H \\
&\qquad\qquad\quad\Big\downarrow{}^{e^-} \\
&\qquad\qquad\qquad\quad H_2O + H
\end{aligned}
\tag{4.3}
$$

Information on a planet's atmosphere is available for transiting planets in comparatively nearby systems: When the planet transits its sun, some of the star's light filters through the planet's atmosphere, thus revealing its composition. Atmospheres thick with clouds have thus been pinned down for the mini-Neptune GJ 1214-b in the constellation Ophiuchus (Kreidberg 2014), as well as for the super-Neptune GJ 436-b in the constellation Leo (Knutson 2014). Both planets orbit their M-type star in the non-habitable zone at very close distances. Data on these two planets and their suns are assembled in Table 4.2. The clouds of GJ 1214-b may be composed of either microscopic droplets of zinc sulfide and potassium chloride, or a haze of hydrocarbons. Similar considerations hold for GJ 436-b: Although the mass of GJ 436-b is about four times that of GJ 1214-b, hinting towards an atmospheric composition similar to that of Neptune, and hence an atmosphere dominated by hydrogen and helium, the data suggest an atmospheric composition with only a few percent of hydrogen/helium mass fraction.

Table 4.2. Selected characteristics of four planets with atmospheres and (for the planets indicated in bold) clouds in their atmosphere

Star/ planet	Mass of planet	Radius of planet	Semi-major axis (AU)	Orbital period (d)	Possible atmospheric constituents	Classifi-cation and age (Gy) of star	Surface temp. (K) of star
GJ 1214						M, 6.0	3026
GJ 1214-b	6.5 $m_\oplus$	2.63 $r_\oplus$	0.014	1.58	KCl, ZnS, or hydrocarbons		
GJ 436						M, 6.0	3684
GJ 436-b	23.4 $m_\oplus$	7.3 $r_\oplus$	0.029	2.64	KCl, ZnS, some H and He		
Kepler 7						(Late F), 3.5	5933
Kepler 7-b	0.44 m_J	1.61 r_J	0.06	4.89	Silicates, e.g., Mg_2SiO_4		
HR 8799						A5, 0.06	-
HR 8799-c	10±3 m_J	1.3 r_J	42.9	82145	H_2O, CO, (CH_4)		

Table 4.2 further contains data for the hot Jupiter Kepler 7-b and the super-Jupiter HR 8799-c (see also Figure 2.3). Kepler 7-b orbits an approximately Sun-like star in the constellation Lyra. The rather large albedo[11] of Kepler 7-b has been attributed to a cloud based on silicates such as Mg_2SiO_4 or $MgSiO_3$ and/or a haze layer originating from condensed hydrocarbons (Demory 2013). The average density of this planet, $\rho = 0.14$ g cm^{-3}, is particularly low, pointing to an extended and thin atmosphere. HR 8799-c is at an average distance of 42.9 AU, and hence far outside its very young and rather hot central star in the constellation Pegasus. The planet's atmosphere mainly contains water and carbon monoxide, and thus substantially differs from that of the three other planets in orbits close to their sun. Despite the ample distance from its sun, the surface temperature of HR 8799-c, ca. 1100 K, is rather high, reflecting the gravitational energy released during the formation of this young planet (Konapacky 2013).

From the point of view of habitability, Earth-like planets, super-Earths and sub-Neptunes (mini-Neptunes) orbiting their sun in a moderate zone are of

---

[11] The albedo of a celestial body is defined by the ratio 'reflected radiation'/'incident radiation' (incident radiation is the radiation received from a planet's sun). An albedo =1 refers to total reflection, and albedo = 0 to zero reflection.

particular interest. An approximately terrestrial planet will melt and differentiate during the early phase of its formation – the accretion phase in the protoplanetary nebula – and also, to some extent, thereafter. The heat for the smelting is provided by the accretion process, by impacts of objects the size of planetesimals, and by radioactive decay of (now "extinct") isotopes with comparatively short half-lives, such as ^{26}Al with a half-life of $7.2 \cdot 10^5$ a; Eq. (4.4).

$$^{26}_{13}\text{Al} \longrightarrow\ ^{26}_{12}\text{Mg} + ^{0}_{1}\beta^+ + ^{0}_{0}\nu$$

$$(4.4)$$

These processes result in the formation of a solid inner and a liquid outer iron-rich core, a silicate-rich dense inner and less dense outer mantle, and a rocky crust + (solid and liquid) water. With increasing mass of the planet, the thickness of the core and the lower mantle increase, while that of the upper mantel decreases.

The recent discovery of the mineral ringwoodite found in a diamond stemming from the transition zone of Earth's mantle, i.e., the transient area between the lower and upper mantle at a depth of ca. 530±120 km, suggests that water is also present in the depths of Earth's mantle, and that such a transition zone of Earth, and Earth-like exoplanets as well, could be a major repository of water (Pearson 2014).

Ringwoodite, a mineral that has previously also been found in meteorites, is a high-pressure polymorph of olivine $(Fe,Mg)SiO_4$, and can contain up to 2.5% by mass of hydroxide, and thus "hidden water". The hydroxide moiety is easily detected by its typical stretching frequencies in the infrared at 3160 and 3580 cm^{-1} (3.165 and 2.793 µm). As ringwoodite is transported to the surface, the accompanying release of the pressure and breakdown of the crystal structure generates water from the hydroxyl groups (2FeOH $\rightarrow$ {FeOFe} + H_2O).

The amount of this burrowed water is about the same as the overall batch of water in the oceans. Iron, magnesium, silicium and oxygen are the dominating elements in Earth's mantle, and also in the rocky part of the outer envelope (the crust), while iron, along with a few percent of nickel, dominates the core.

4.3. METEORITES AS GUIDES TO THE ORIGIN OF PLANETARY MATERIAL

Most meteorites originate from interplanetary fragments that formed in the infancy of the Solar System, and thus hold fingerprints from the early days. Others are fragments that were generated by disruption of an asteroid or detachment of planetary material in the wake of instability (in the case of asteroids) or a major impact (in the case of asteroids and planets), commonly at a comparatively early stage of the constitution of our system. The planetesimals or asteroidal fragments can be composed of rocky and/or metallic (essentially iron) material, and can incorporate appreciable amounts of icy components and of regolith, i.e., more or less finely grained bedrock. Comets can also be sources of meteorites. About 150 meteorites of Martian origin have so far been identified; the vast majority of meteorites are, however, of pre-planetary or asteroidal origin. A rocky or metallic fragment of a planetesimal or an asteroid captured in Earth's field of gravity is termed a "meteoroide"; as the meteoroide hits the atmosphere and starts to glow and burn, producing a light streak, it is called "meteor". When sufficiently large and compact, parts of it survive the passage through Earth's atmosphere, and the remainders can be collected as meteorites. Their physical, mineralogical and chemical properties provide valuable hints in respect to their origin.

The bulk composition of Earth – in particular the abundance of specific isotopes of nitrogen, oxygen, titanium, chromium and nickel – corresponds to that of meteorites belonging to the so-called EH chondrites, where E stands for *e*nstatite (magnesium silicate $MgSiO_3$), and H for *h*igh contents of iron. The term *chondrite*[12] refers to the presence (in most cases) of *chondrules* – micro- to millimeter-sized more or less globular inclusions of originally molten on mineral ingredients, and commonly in an envelope of agglomerated dust (see Figure 4.3). Chondrules are rich in the minerals olivine and pyroxene. Olivine is a silicate of composition $(Fe,Mg)_2SiO_4$, containing varying amounts of iron and magnesium. Pyroxenes are complex alumosilicates of metals such as Fe, Mg, Co and Mn. Feldspar and zeolites are examples of alumosilicates. EH chondrites make up only a comparatively small fraction of the large family of chondrites. Chondrites are essentially stony meteoritic objects stemming from the time when accretion started in the presolar nebula; they have experienced

---

[12] Achondrites are stony meteorites *without* chondrules. They originate from asteroids, and have experienced substantial transformation by melting since their formation.

little change during their original formation, and hence carry features and thus information from the dawn of our home system. Along with these mineral chondrules, chondritic meteorites commonly contain flecks that are due to iron-nickel embeddings. Enstatites are most chemically reduced; their iron nickel contents (in the form of metallic Fe-Ni as well as sulfides and silicides of Fe and Ni) can go up to 10% of the overall mass.

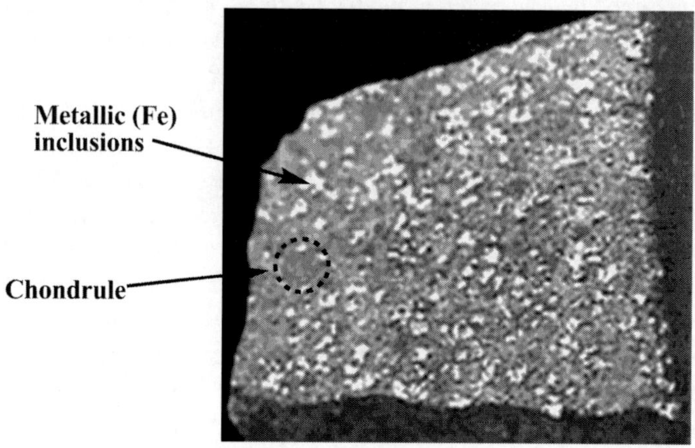

Figure 4.3. A fragment from a meteorite (so-called Hammami meteorite, named for the find spot in the El Hammami mountains in Mauretania) belonging to a subgroup of common chondrites, the H (*h*igh iron) chondrites. Reproduced with permission from Michael Farmer meteorites.

A second class of chondrites is known as *common chondrites*. Common chondrites comprise three closely related groups, named H chondrites (*h*igh iron), L chondrites (*l*ow iron), and LL chondrites (*l*ow iron and *l*ow overall metal content). Common chondrites often have comparatively large chondrules (Figure 4.3) of a complex structure and embedded in several layers formed by the dust present in the realm of their origin. The chemical, physical and mineralogical properties of chondrites, along with the isotope pattern, shed light on the environment and the period of time in which the chondrules formed, and thus allow for deductions with respect to the area in the presolar nebula where these objects originally accrued (Rubin 2010): Common (H, L and LL) chondrites are associated with the asteroid belt, from where they can be slung by the tidal forces exerted by the giant planet Jupiter towards the inner planets. In contrast, enstatite (E) chondrites derive from a region approximately stretching between today's orbits of Venus and Mars.

A third and rare class of chondrites, the *Rumuruti-* or R chondrites (named for the location of their fall in Kenya), originates from the outer zone (around 3 AU) of the asteroid belt. R chondrites are characterized by a comparatively high amount of the oxygen isotope ^{17}O relative to the common isotope ^{16}O. ^{16}O is a *primary* isotope formed in a star in the frame of the so-called 3α process, Eq. (4.5), while ^{17}O is a secondary isotope generated by, e.g., proton capture through ^{16}O and subsequent β^+ decay of the intermittently formed ^{17}F (Eq. (4.6); cf. also sidebar 2.1). Levelling of the isotopic ratios indicates an origin from a cooler region, thus more distant from the Sun.

$$\ce{^{4}_{2}He + ^{4}_{2}He ->[\gamma] ^{8}_{4}Be ->[^{4}_{2}He][\gamma] ^{12}_{6}C ->[^{4}_{2}He] ^{16}_{8}O + \gamma} \tag{4.5}$$

$$\ce{^{16}_{8}O + ^{1}_{1}H ->[\gamma] ^{17}_{9}F -> ^{17}_{8}O + ^{0}_{1}\beta^+ + \nu} \tag{4.6}$$

Still another group of chondrites, originating from the outer region of the asteroid belt and beyond, is represented by the *carbonaceous* chondrites (C chondrites). These are chondrites with up to 3% carbon contents in the form of carbonate, carbide, organic carbon compounds (kerogens) and elemental carbon, and thus compounds that cannot survive the high temperature and irradiation in areas closer to the protosun. With the exception of the enstatite chondrites (with rather tiny chondrules) the buildup of the chondrites clearly hints towards their formation in the dense dusty area of the protoplanetary system. This dust area extended from ca. 2.7 and 4.5 AU, and hence beyond today's outer margin (3.3 AU) of the asteroid belt. We will refer back to carbonaceous chondrites in section 6.2.

SUMMARY

An appreciable number of exoplanets have radii between 1 and 4 $r_\oplus$, and thus range in size between Earth, super-Earths (up to ~2 $r_\oplus$), and sub- or mini-Neptunes (below ~4 $r_\oplus$). These planets are of particular interest with respect to compositional features associated with potential habitability. Planets ranging in size between Earth and ca. 2 $r_\oplus$ have an iron-dominated core and a mostly rocky mantle, with overall densities above 5 g cm^{-3} (Earth: 5.51 g cm^{-3}). These rocky planets are thought to be especially suited as abodes for life. Larger planets contain a pronounced fraction of volatiles such as atomic hydrogen, helium and water (Marcy 2014). Consequently, mean densities of planets generally decline with increasing planetary radius. An appreciable number of the known super-Earths and sub-Neptunes orbit their sun at a distance < 0.25 AU (and thus commonly in the non-habitable zone). The abundance of this genre of planets reflects the feasibility of detection (by the transit method or by Doppler shift) of planets in nearby orbits, rather than the actual distance distribution of planets in extrasolar systems.

While the size of an exoplanet is comparatively easily ascertainable, the mass and density and, thereof derived, the composition of an exoplanet is often inaccessible or only vaguely available. The metallicity of the host star can provide a first clue to a planet's buildup and composition. Transit timing variation and ultra-high precision Doppler spectroscopy have lately advanced the status of the respective information for planets ranging in size between super-Earths and mini-Neptunes. Such planets supposedly have an iron core, a rocky mantle covered by water (ice), and an extended atmosphere dominated by H_2 and He. In a few approximately Neptune-sized planets orbiting their central star at a close distance, dense clouds have been detected that consist of droplets of potassium chloride and zinc sulfide, plus siliceous dust.

Jupiter-sized planets have atmospheres reminiscent of Jupiter and Saturn in our system, and hence with hydrogen and helium as the main constituents. Depending on the distance to their sun, minor components come in, including H_2O, CO, CH_4, H_2S and the hydrides of P, As and Ge in *distant* Jupiters, vs. O, H_2O, CO_2 and CH_4 in *hot* Jupiters.

The differentiation inside Earth-like planets and super-Earths takes place in an early stage of their formation. The heat necessary for the melting process allowing for this differentiation is provided during planetary accretion, by impacts, and by the decay of short-lived radioactive nuclei. Water is in part incorporated during the accretion phase, and likely "stored" by minerals related to olivine in the transition zone between the lower and the upper mantle. At later stages of the planetary development, water is delivered by comets and meteorites.

In the Solar System, the composition, as well as the micro- and macroscopic appearance of meteorites (chondrites in particular) provide a valuable tool for judging the intrinsic properties of presolar nebulae and the specific situation in our system at an early stage of its constitution, such as the composition and texture of dust and grain, the radiation intensity of the protosun, and weathering by cosmic dust and cosmic rays as a function of the distance to the Sun. These are features that can be extrapolated to comparable situations in extrasolar systems.

REFERENCES

Borucki, W.J., E. Agol, F. Fressin, et al. (2013): Kepler-62: A five-planet system with planets of 1.4 and 1.6 Earth radii in the habitable zone. *Science* 340:587-590.

Buchhave, L.A., M. Bizzarro, D.W. Latham, et al. (2014): Three regimes of extrasolar planet radius inferred from host star metallicities. *Nature* 509:593-596.

Crossfield, I.J.M., B. Biller, J.E. Schlieder et al. (2014): A global cloud map of the nearest known brown dwarf. *Nature* 505:654-656.

Demory, B.-O., J. de Wit, N. Lewis, et al. (2013): Inference of inhomogeneous clouds in an exoplanet atmosphere. *Astrophys. J. Lett.* 776:L25.

Fressin, F., G. Torres, J.F. Rowe, et al. (2012): Two Earth-sized planets orbiting Kepler-20. *Nature* 482:195-198.

Howard A.W. (2013a) Observed properties of extrasolar planets. *Science* 340:572-576.

Howard, A.W., R. Sanche-Ojeda, G.W. Marcy, et al. (2013b): A rocky composition for an Earth-sized exoplanet. *Nature* 503:381-384.

Kipping, D.M., D. Nesvorný, L.A. Buchhave, et al. (2014): The hunt for exomoons with Kepler (HEK): IV. A search for moons around eight M-dwarfs. *Astrophys. J.* 784:28-41. See also Y. Bhattacharjee (2014): Star-crossing planets literally strut their stuff. *Science* 343:240.

Knutson, H.A., B. Benneke, D. Deming, et al. (2014): A featureless transmission spectrum for the Neptune-mass exoplanet GJ 436b. *Nature* 505:66-68.

Konopaxky, Q.M., T.S. Barman, B.A. Macintosh, et al. (2013): Detection of carbon monoxide and water absorption lines in an exoplanet atmosphere. *Science* 339:1398-1401.

Kreidberg, L., J.L. Bean, J.-M. Désert, et al. (2014): Clouds in the atmosphere of the super-Earth exoplanet GJ 1214b. *Nature* 505: 69-72.

Marcy, G.W., H. Isaacson, A.W. Howard, et al. (2014): Masses, radii, and orbits of small *Kepler* planets: the transition from gaseous to rocky planets. *Astrophys J. Suppl. Ser.* 210:20.

Miguel, Y., and L. Kaltenegger (2014): Exploring atmospheres of hot mini-Neptunes and extrasolar giant planets orbiting different stars with applications to HD 97658B, WASP-12B, COROT-2B, XO-1B and HD 189733B. *Astrophys. J.* 780:166.

Pearson D.G., F.E. Brenker, F. Nestola, et al. (2014): Hydrous mantle transition zone indicated by ringwoodite included within a diamond. *Nature* 507:221-224.

Quintana, E.V., T. Barclay, S.N. Raymond, et al. (2014): An Earth-sized planet in the habitable zone of a cool star. *Science* 344, 277-280.

Rubin, A. (2010): Physical properties of chondrules in different chondrites groups: Implications for multiple melting events in dusty environments. *Geochim. Cosmochim. Acta* 74:4807-4828.

Chapter 5

HOW TO DEFINE LIFE

In the light of what has been set out in the preceding section 4.3, carbonaceous meteorites are of primary interest with respect to their contents of carbon-based molecules, and organic molecules in particular, when it comes to the conception of Life as we know it. Carbonaceous meteorites present just *one* extraterrestrial source in the potential supply of compounds that can be starter composites or building blocks for life molecules on Earth as well as on planets in other systems: in sections 3.1 and 3.2, and in sidebar 3.1, we have already pointed to additional repositories, such as rogue planets, comets, dense interstellar clouds of gas and dust, and matter ejected into space by maturing and by dying stars. The dust and gas clouds of planetary nebulae are examples of the latter. Exoplanets in the *habitable* zone, in particular those that resemble Earth in size and composition, are particularly intriguing sources for an origin of life. In reference to extremophiles on our home planet, the prevalent perception of habitable needs, however, to be challenged and amplified to some degree.

What are the implications of the term Life? To a certain extent, this is clearly a question that alludes to a philosophical dimension. But even when disregarding philosophical aspects, the answer to this question is far from being simple, and this issue is thus being disputed in science. Is, by way of example, a virus a living organism? Another question in this context concerns the necessity of water for the sustenance of life. As mentioned repeatedly in the previous chapters, we are accustomed to the notion that, without water, life cannot exist; and even more so that live-giving and life-sustaining water has to be present in the liquid form, restricting the temperature of potential habitats roughly to the range defined by water's freezing and boiling points. In addition to the water issue, we will also critically scrutinize the possibility of molecules of life that do not depend on carbon, and thus on organic compounds that are indispensable in exclusively all of our terrestrial life forms.

In this chapter, I will coin a pragmatic view for the conception Life, discuss and validate potential habitats that are devoid of water, and challenge the notion of the "carbon world" as a prerequisite for life. At the end, I will put the reader's mind at rest – as far as water and carbon are concerned, but not with respect to the *origin* of Life; this issue will be intensified in Chapter 6.

5.1. ELEMENTS AND MOLECULES IN LIVING ORGANISMS

5.1.1. Carbon-Based Life Molecules

Living organisms thriving on our home planet are commonly classified into three kingdoms of life, namely bacteria, archaea and eukarya. Bacteria and archaea are also summed up under the term prokarya, meaning that they are devoid of a nucleus. (For a more detailed characterization of eukarya and prokarya see further down.) All organisms on our planet are believed to have evolved from a last uniform common ancestor, LUCA, about 3.6 Ga ago (Plaxco and Gross, 2011) (Figure 5.1). Life on Earth as still represented by prokarya in their present appearance likely started about 3.5 Ga ago, i.e., ca. one billion years[13] after the consolidation of our planet in its present state and position in the Solar System – provided that life became initiated on Earth at all; see section 6.3 for potential alternative sources. These signs of ancient life are attested by, *inter alia*, microbially induced 3.48 Ga old sedimentary structures found in the Dresser Formation in Western Australia (Noffke 2013) (see also the time-line, Figure 5.10, in section 5.4).

Let us, to begin with, assume that organic macro-molecules serve as a basis for life. Such organic building blocks are provided by proteins, lipids, carbohydrates (sugars, starch, cellulose, chitin); and by molecules responsible for information storage, the expression of information, and information transfer. These information bearing organic polymers are nucleic acids, abbreviated as RNA (ribonucleic acids) and DNA (deoxyribonucleic acids). In

---

[13] 1 Ga may be a somewhat optimistic assumption for the time span available for the development of life: 4.1-3.8 Ga ago, our planet was subjected to a cataclysmic "heavy bombardment" by asteroids, leaving just a few hundred thousand years – after the cessation of the bombardment – for life to come into existence.

the following, some basic insight into the chemical nature and life function of these four groups of complex molecules will briefly be provided, followed by the introduction of examples for "artificial" molecules that might substitute DNA as information carriers. Finally, the issue of whether viruses can be placed within the concept of "life" will be scrutinized.

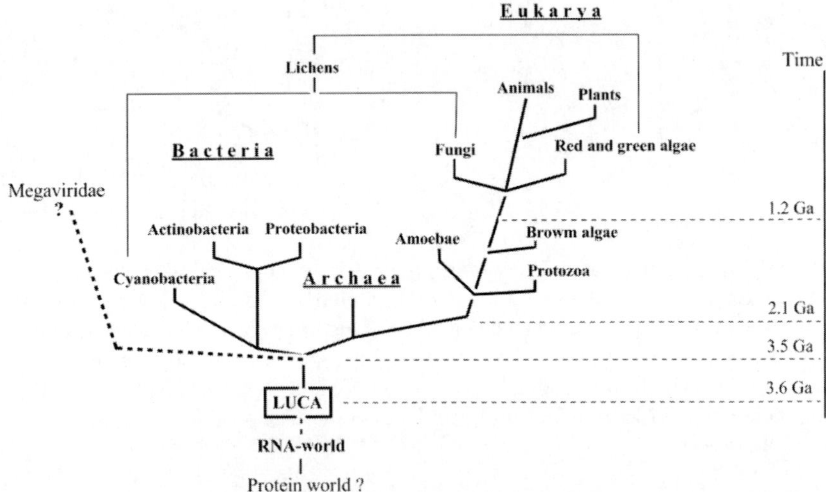

Figure 5.1. The phylogenetic tree shows the three kingdoms of life (bacteria, archaea and eukarya) as well as selected groups of organisms. Lichens are symbionts between fungi and algae or cyanobacteria. Time intervals are indicated in units of billion years (Ga). LUCA stands for the Last Uniform Common (or Cellular) Ancestor. The RNA world is a potential pre-life form, and might have been preceded by a protein world. According to a recent suggestion, peptide catalysts (so-called urzymes) co-evolved with the RNA, substantially accelerating reactions to translate the genetic code (Li 2013). Provisionally included is a fourth branch, megaviridae. For justification, see the discussion on the viability of viruses at the end of section 5.2.

Proteins are macro-molecules built up of amino acids. Amino acids in turn are carbonic acids with an amino ($-NH_2$) group, commonly attached – in living matter – to the α-carbon (Cα) of the carbonic acid; Figure 5.2. Cα is the carbon next to the carbonic acid group ($-CO_2H$). Two or more amino acids can be linked together by condensation (the elimination of water) and concomitant formation of a C–N linkage, the peptide bond. In the case of two amino acids, the result is a dipeptide (see Figure 5.2c). As more amino acids combine, oligopeptides and finally proteins form.

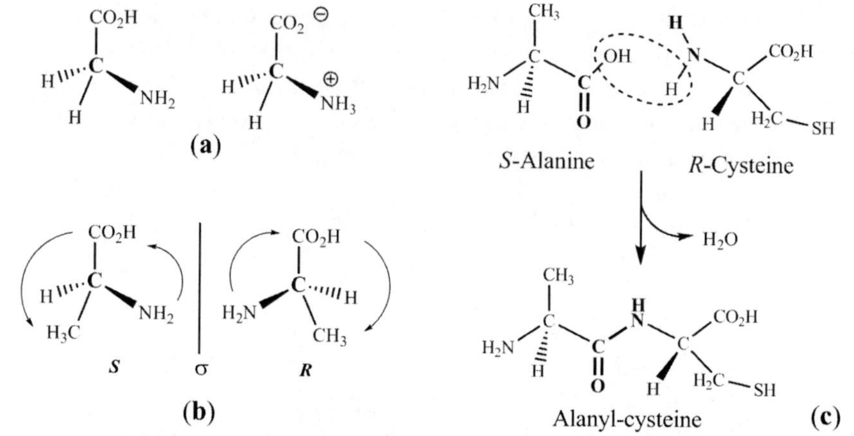

Figure 5.2. (a) Glycine is the only essential amino acid without a chiral center. Shown are the neutral and the zwitterionic form. (b) The amino acids *S*- (or L-) and *R*- (or D-) alanine. The direction of rotation (starting with the highest priority group) is indicated by bent arrows, the mirror plane by σ. The central, tetrahedrally configurated chiral carbon atom (Cα) is shown in bold. (c) Formation of the dipeptide alanyl-cysteine by condensation of *S*- (or L-)alanine and *R*- (or L-)cysteine. The peptide group is highlighted in bold.

With the exception of the most simple amino acid, glycine (Figure 5.2a), the carbon bearing the NH_2 group is asymmetric[14], or chiral, meaning that the carbon atom carries four *different* substituents that can be arranged left-handed or right-handed. This is exemplified by the amino acid alanine in Figure 5.2b. The two stereo-isomers are called enantiomers, and designated *S* (for *sinister* = left) and *R* (for *rectus* = right), where *S* and *R* indicate the direction of rotation, i.e., clockwise (*R*) and counter-clockwise (*S*).[15] 'Direction of rotation' refers to the sequence in which the substituents on the chiral carbon are arranged; starting point is the substituent with the 'highest priority', i.e., the highest atomic number of the atom directly binding to the central carbon. For the alanine molecule shown in Figure 5.2b this is the amino group (^{14}N vs. ^{12}C). Next is the substituent with the second highest priority. As two substituents

---

[14] There is no symmetry operation (except for the unit operation, or identity) by which the molecule can be transformed into itself. Examples for symmetry operations are rotation by a fixed angle (e.g., 60° or 90°), reflection at a mirror plane, and point reflection (inversion).

[15] Alternatively, the absolute configurations L (laevus = left) and D (dexter = right) are in use. "Absolute" refers to the designation of configurations in analogy to glyceraldehyde as a reference; cf. Fig. 5.3. For most amino acids *S* corresponds to L, and *R* to D.

bind via the same atom, as in the case of $-CO_2H$ and $-CH_3$, the second sphere atom of the substituent determines the priority in that the heavier atom again comes first; hence, carboxylate > methyl. The two enantiomers are related to each other such as image and (non-superposable) mirror image. At the very most, amino acids involved in life molecules are the *S* enantiomers.[16]

Lipids are involved in energy storage, signaling and the buildup of cell membranes. Typical lipids that function as structural components for cell membranes derive from the *tris*alcohol glycerol. A specific group of lipids, the phospholiphids, are constituents of cellular membranes. Here, one of the terminal alcoholic groups is esterified with phosphate. Other features depend on the kingdom of life (cf. Fig. 5.1) to which the respective organism belongs:

- *Eukarya* are uni- and multicellular organisms, the differentiated cells of which are compartmentalized. The cells further have available a nucleus with large linear chromosomes (containing the DNA) and mitochondria, the "steering centers" of the cell. Amoebae, protozoa, fungi, plants and animals belong to the eukaryan kingdom. The lipid bilayer constituting the cellular membrane derives from D-glycerol, with two of the alcoholic groups esterified with fatty acids such as linolenic and oleic acid; Fig. 5.3a.
- *Bacteria* are unicellular organisms devoid of compartmentalization and a nucleus, and with comparatively small about circular chromosomes. The buildup of the cell membrane is very much the same as in eukarya.
- *Archaea* again are unicellular organisms without any compartmentalization. Contrasting bacteria (and eukarya), the cell membrane is a lipid *mono*-layer. The lipid derives from L-glycerol (instead of D-glycerol), with two of the alcoholic groups etherified (instead of esterified); see Fig. 5.3b.

Some eukaryan cellular constituents, chloroplasts and mitochondria, in particular, are believed to originate from archaea. Note that viruses have *not* been included here as life forms. For a justification of this exclusion on the one hand, and arguments that relativize this debarment on the other hand, see section 5.2.

---

[16] As a consequence of the higher priority of the $-CH_2SH$ vs. the $-CO_2H$ substituent, cysteine is present in the *R* configuration. The *absolute* configuration (cf. footnote 15) of naturally occurring cysteine is the L form, hence L-cysteine.

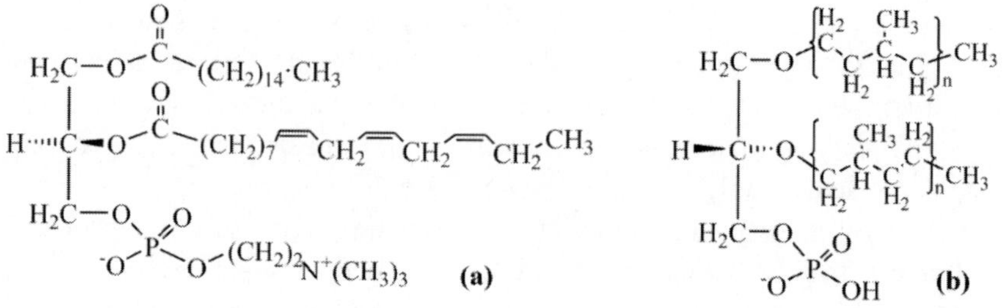

Figure 5.3. Lipids that constitute the cellular membrane of (**a**) eukarya and bacteria, and (**b**) archaea. The central carbon in the glycerol moiety is in the L configuration in (**a**), and in the D configuration in (**b**). The carbonic acid substituents in (**a**) derive from hexanoic acid (upper end) and linoleic acid (central part); the phosphate group is additionally esterified with choline. The carbon chain fragments cased in curly brackets in structure (**b**) are isoprene units.

Carbohydrates serve as constituents in storage molecules for energy, for example in polysaccharides such as starch, and as building blocks and structural components, for example in cellulose (plants) and chitin (fungi, arthropods and crustaceans). The building blocks for the polymers starch and cellulose are glucose molecules; glucose is also one of the two components of cane sugar and beet sugar; the second component is fructose. Chitin is a polymer that is composed of acetyl glucosamine, and hence a derivative of glucose where one of the hydroxyl functions is replaced by an amino group carrying an acetyl (CH_3CO) substituent. Figure 5.4 provides details of the formulae.

Also shown in Figure 5.4 are ribose and deoxyribose. These two sugar molecules are components of *ribonucleic acid* (RNA) and *deoxyribonucleic acid* (DNA), respectively. The last universal common ancestor LUCA (cf. Figure 5.2) already employed DNA as an information *carrier*, as do prokarya (archaea and bacteria) and eukarya while, in the RNA world proposed to have existed prior to LUCA, the (less stable) RNA was utilized. Viruses can apply DNA *or* RNA to store information, depending on the specific virus under consideration.

Both prokarya and eukarya use RNA in information *transfer*, i.e., in the transcription of information (stored in DNA) into protein settings. Finally, RNA can also be a biological catalyst for the synthesis of proteins from amino acids, and is hence comparable to protein-based enzymes. These RNA-based enzymes, termed ribozymes, may also have played a role in the putative RNA world.

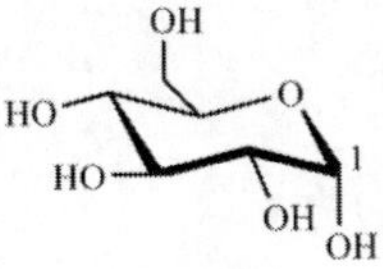

α-D-Glucose

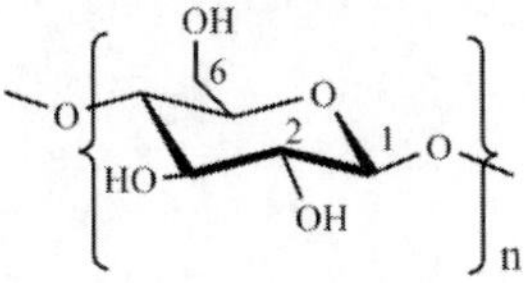

Cellulose = (β-D-glucose)$_n$

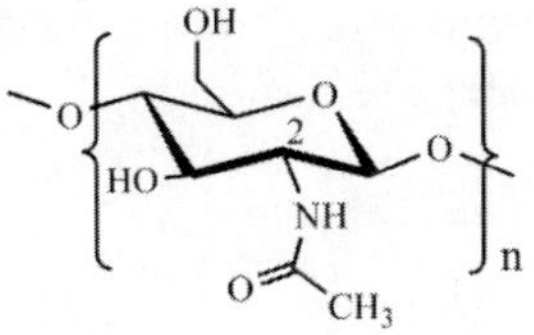

Chitin =
(β-acetylglucosamine)$_n$

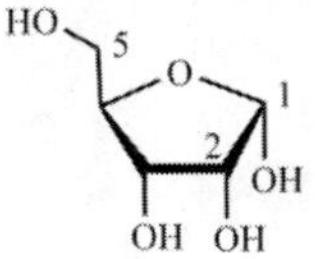

α-D-Ribose

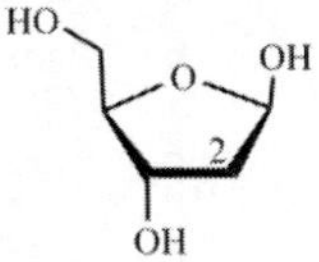

2-Deoxy-β-D-ribose

Figure 5.4. Sugars that play a role in energy storage and as structural components (upper row), and as constituents of ribonucleic acids (ribose) and 2-deoxyribonucleic acids (deoxyribose). The descriptors α and β refer to the axial (α) and equatorial (β) position, respectively, of the O(H) group linked to the carbon atom C1.

While DNA forms double-stranded helices, RNA is usually single-stranded. The monomeric units of these nucleic acids, the nucleotides, contain a sugar (ribose in the case of RNA, deoxyribose in the case of DNA), phosphate, and a so-called nucleobase. In the case of DNA, the nucleobases are the (complementary pairs) adenine + thymine and guanine + cytosine. In RNA, thymine is replaced by uracil. Nucleobases are nitrogen containing six-membered or 6+5-membered cyclic compounds. Details are outlined in (a)-(c) in sidebar 5.1.

***

Sidebar 5.1. The constitution of genuine RNA and DNA, and artificial alternatives thereof

Information storage and processing in virtually all life forms on our planet exclusively relies on the nucleic acids DNA and RNA. Does the restricted number of building blocks employed for these polymeric molecules in exclusively *all* life forms reflect fundamental constraints with respect to their genesis in the primordial broth, or a coincidental evolutionary track? Recent lab-made alternatives of DNAs in fact provide evidence that storage and transmission of information do not necessarily rely on the nucleic acids that dominate terrestrial life.

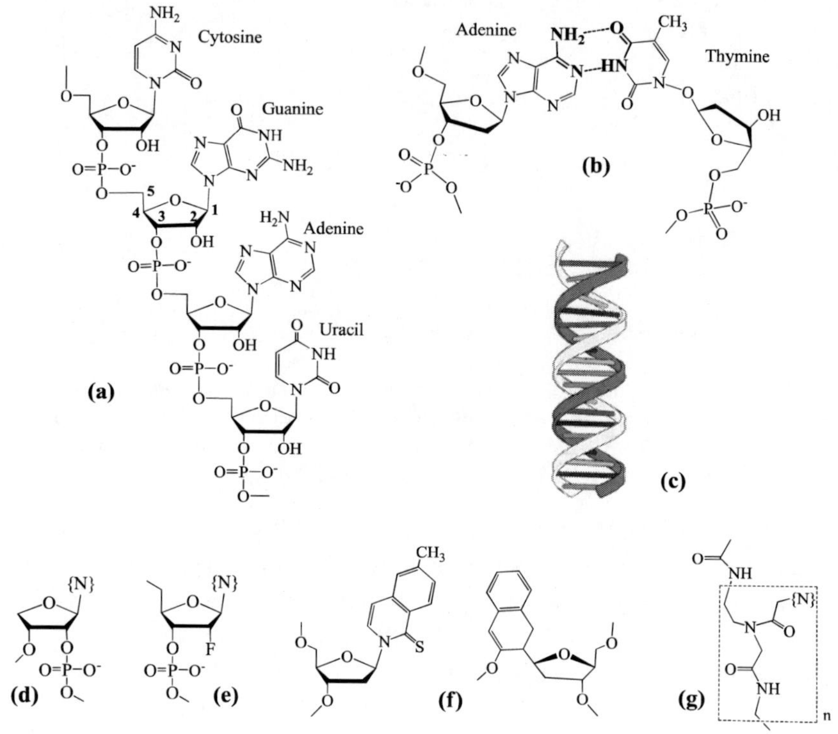

(**a-c**) Genuine ribonucleic and deoxyribonucleic acids: The (in the majority of cases) single-stranded ribonucleic acids RNA (**a**) contain the four nucleobases cytosine and uracil (pyrimidine bases), and guanine and adenine (purine bases). The bases are linked to the C1 of ribose; ribose in turn links to bridging phosphates via C3 and C5. The ribose-phosphate chain forms the back-bone of the polymeric structure. In deoxyribonucleic acids DNA (**b**), ribose is replaced by deoxyribose. In deoxyribose, the OH group at C2 is missing; see also the bottom part in Fig. 5.4. As with RNA, DNA contains the nucleobases cytosine, guanine and adenine; uracil in RNA is replaced by thymine in DNA. DNA is mostly double-stranded, with the two strands forming left-handed helices (**c**). The bases are aligned along the common axis of the two strands. Two facing bases in the helical strands are connected via hydrogen bonds, as shown in (**b**) for the pair adenine-thymine. In (**c**), these hydrogen bonding interactions are symbolized by horizontal bars connecting the two strands.

(**d-f**) Artificial, or xeno-nucleic acids XNA, are assembled correspondingly. In (**d**) and (**e**), the sugars are no longer ribose/deoxyribose, but threofuranose (**d**) and 2-fluoroarabinose (**e**). The unnatural "base pair" (**f**) has been successfully incorporated into the DNA of *Escherichia coli*. Here, base pairing is accomplished by π interaction between the aromatic moieties rather than by hydrogen bonding as in (**b**). For references see the main text.

(**g**) In the peptide-nucleic acids PNA, the backbone is no longer composed of sugar-phosphates as in common RNA/DNA and XNA, but of a peptide chain instead. {N} represents a purine/pyrimidine base.

***

Alternate DNA molecules with sugars other than deoxyribose, or bases other than adenine, thymine, guanine, cytosine and uracil have been shown to store and recover genetic information. These alternate DNAs are commonly referred to as xeno-nucleic acids XNA. Examples are XNAs with X = threofuranose (Schönig 2000) or 2'-fluoro-arabinose (Pinheiro 2012), and XNAs with artificial nucleobases.

The respective sugars and bases are illustrated in the bottom line in sidebar 5.1, (d) and (e). The artificial base pair (f) has recently been incorporated into the *E. coli* DNA (Malyshev 2014). In principle, such a DNA with *three* base pairs could allow for the information storages of a total of 172 amino acids, rather than the 21 essential amino acids[17] encoded by the "classical" DNA with just *two* base pairs. An even more 'radical' approach to alternatives of DNA are peptide nucleic acids PNA, (g) in sidebar 5.1. Here, the ribose-diphosphate backbone is replaced by a peptide backbone (Nielsen 1999); (g) see sidebar 5.1.

So far, we have defined simple and complex carbon-based molecules that form the basis of life as we know it. We will outline in Chapter 6 the primordial generation and supply of these molecules. Prior to this, we will address the conundrum of "Life" as such, i.e., we will set out criteria that allow us to distinguish between animate and inanimate molecular assemblies. Furthermore, we will briefly approach the question of whether (i) life can

---

[17] Including selenocysteine, hence 21 instead of the commonly quoted 20 essential amino acids.

resort to non-carbon molecules, and (ii) life is restricted to the operational availability of water.

5.1.2. Inorganics in Life

Several of the molecules dealt with in the preceding section – lipids, DNA and RNA – are activated by phosphate, an anionic inorganic component derived from *ortho*-phosphoric acid H_3PO_4. At physiological pH (commonly close to pH 7), phosphate is essentially present in the form of mono- and di-hydrogenphosphate, HPO_4^{2-} and $H_2PO_4^{-}$. Phosphate, in the form of hydroxyapatite $Ca_5(PO_4)_3OH$ and fluor-hydroxyapatite $Ca_5(PO_4)_3(OH)_xF_{1-x}$, is significantly involved in bone structures. Inorganic phosphate is, however, also an indispensable constituent in a plethora of organic (i.e., carbon-based) functional molecules and thus intimately involved in life processes. In addition to lipids (Figure 5.3), DNA and RNA (sidebar 5.1), various enzymes rely on phosphate as an "activator".

Examples are the insulin receptor, trans-membrane transport systems for metal ions such as Na^+ and K^+, and hydrolases – enzymes that catalyze the splitting of an (ester) bond with the help of water. Hydrogenphosphate can condense, forming diphosphate and triphosphate. The latter is an essential constituent of adenosine-triphosphate, ATP, the Mg^{2+} activated hydrolysis of which (Figure 5.5) delivers energy.

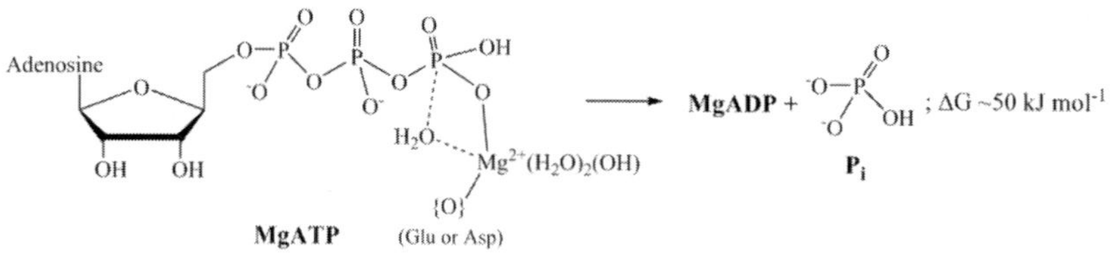

Figure 5.5. The hydrolysis of Mg^{2+}-ATP (ATP = adenosine triphosphate) delivers Mg^{2+}ADP (ADP = adenosine diphosphate) and P_i (inorganic phosphate). The Gibbs free reaction energy is ca. 50 kJ mol^{-1}. For adenosine see sidebar 5.1. Glu and Asp (glutamate and aspartate, symbolized by {O}) are coordinated to Mg^{2+} via the carboxylate-O.

Apart from magnesium's role in ATP hydrolysis, the presence of Mg^{2+} in chlorophyll is noteworthy. Mg^{2+} further triggers various activation paths. Other inorganic elements that are essential for life encompass many metal and non-metal ions. Figure 5.6 provides an overview.

The alkaline metal ions Na^+ and K^+ are involved in the regulation of the osmotic pressure and membrane potentials. In multicellular organisms, Na^+ is the main extracellular ion, while K^+ is the main intracellular ion. The redox-active transition metals manganese, iron, copper and molybdenum are present in the active center(s) of a broad variety of enzymes that are involved in physiologically relevant redox reaction.

Cobalt is a constituent of vitamin-B_{12}. Zinc, next to iron the most abundant transition metal in living organisms, is present in hydrolases and carboanhydrase (the enzyme that catalyzes the interconversion of carbonic acid and water + carbon dioxide), and takes over a crucial role in genetic transcription. The role of fluorine has already been addressed in the context of bone structures.

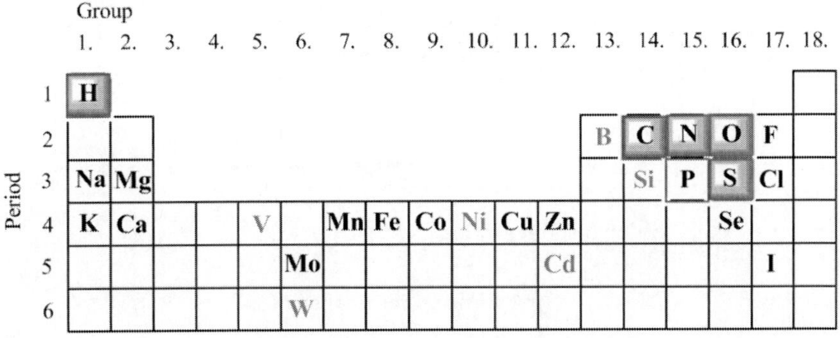

Figure 5.6. Periodic table of the life elements. Elements essential for all organisms are shown in black; elements that are essential for just a limited group of organisms in grey. Elements constituting basic organic matter (H, C, N, O and S) are highlighted by grey boxes. In addition, these "organic elements" are also present in functional inorganic molecules. Examples are hydrogen carbonate HCO_3^- in the regulation of the blood serum pH, NO, CN^- and H_2S in signal transduction, and sulfate SO_4^{2-} as the final product of the degradation of sulfur-based amino acids.

Chloride is, along with HCO_3^-, HPO_4^{2-} and SO_4^{2-}, an important anion in the intra- and (predominantly) extra-cellular space. Iodide is a constituent of the phylogenetically very old thyroid hormones, and selenium is an integral part of the essential amino acid selenocysteine. For a more detailed presentation of the role of inorganics in life see sidebar 5.2.

***

Sidebar 5.2. Overview of the role of inorganics in life

The following scheme provides a general – though not complete – overview (Rehder 2014) of the general biological functions of essential transition and non-transition metal ions (see also Figure 5.6) in living organisms.

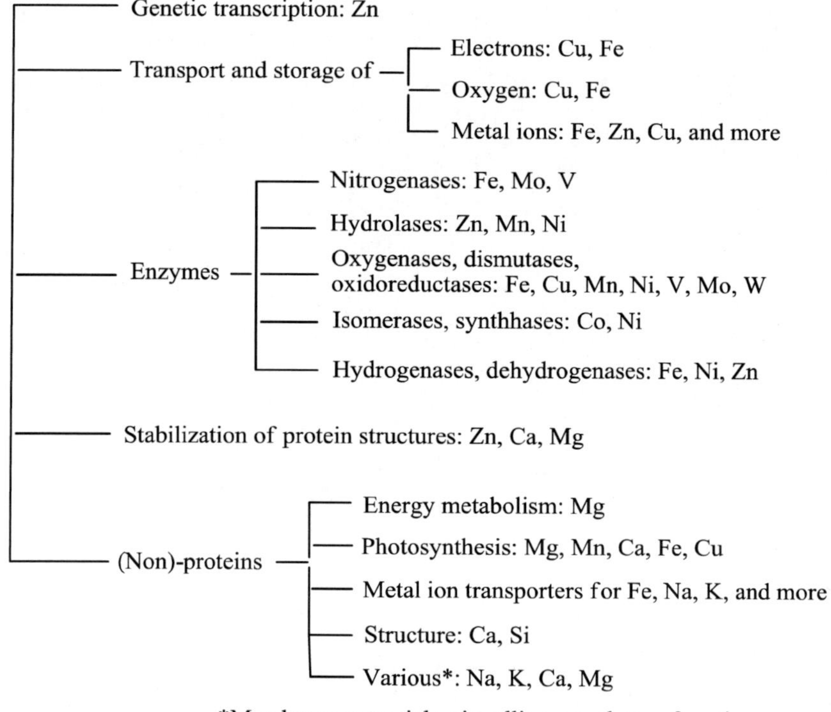

A briefing on selected functions of metalloids and non-metals is provided below, stressing the significance in inorganic processes that are important in physiology:

- Boron, in the form of borate, stabilizes herbal cell walls by crosslinking pectic polysaccharides.
- Carbon, inorganic: Hydrogen carbonate HCO_3^- (formerly known as bicarbonate) is the main buffer in blood serum, and further involved in the off-transport of CO_2. Carbon monoxide CO and cyanide CN^-,

apart from being highly toxic at non-physiological concentrations, exert signaling functions.

– Chlorine, in its ionic form chloride Cl^-, is the main extracellular anion, regulating ion homeostasis and electrical excitability.

– Fluorine: Fluoride F^- partially replaces OH^- in apatite. In human bone, ca. 0.01% of hydroxide is substituted by fluoride.

– Iodine is an essential constituent in thyroid hormones.

– Nitrogen is, along with H, C, O, and S, an omnipresent element in organic biomolecules. Inorganic nitrogen compounds such as nitrate NO_3^-, nitrite NO_2^-, ammonia/ammonium ions NH_3/NH_4^+, hydrazine N_2H_4, nitric oxide NO and nitrous oxide N_2O play a central role in the microbial nitrogen metabolism.

– Phosphorus, in the form of inorganic phosphate, is a constituent in the bone structure. Esters of phosphate are involved in a plethora of energetically conducted reactions, in hydride transfer agents, in lipids, DNA and RNA. For details, see section 5.1.2, sidebar 5.1 and Fig. 5.3.

– Selenium is a constituent in selenocysteine (S in cysteine is substituted for Se).

– Silicon: Silicates participate in the buildup of nascent bone structures, silica gels $SiO_2 \cdot xH_2O$ in the stabilization of structures in grass, horsetail (*Equisetum*), and the shells of diatoms.

– Sulfur is present in the S-functional amino acids, cysteine, cystine, homocysteine, methionine and S-methylmethionine. Inorganic sulfur, in the form of sulfate, is a constituent in the extracellular fluids, and can also link to, for example, sugars, forming sulfones $HOSO_2(OR)$. The generally toxic hydrogensulfide H_2S takes over signaling tasks at very low concentrations. Several sulfur bacteria contain elemental sulfur in the form of cyclic $\rho\text{-}S_6$.

***

5.2. What Is Life? – Metabolism, Reproduction, Evolution, and Communication

As noted previously, to define Life is a complex task, if not a challenge bearing even a philosophical dimension. Life is, of course, not just lipids, carbohydrates and nucleobases, plus phosphate and other inorganics that sustain life processes. A dead individual, although still endued – shortly after its demise – with the same set of chemical ingredients as its living counterpart, does not fulfill all of the criteria for life. On a rather superficial level, we may denote obvious manifestations of life such as growth, reproduction (self-replication, to a large extent), metabolism, and evolution as criteria for life. And in fact they are; though not when considered as isolated criteria. Crystals of sugar and salt (sodium chloride), for example, *grow* in a saturated aqueous sugar or salt solution, and they even *reproduce*: as a sugar crystal is injected into a supersaturated aqueous solution of sugar, this crystal initiates the "birth" and growth of other sugar crystals. Also, "worms" in cyberspace are capable of self-replication. Examples of infectious organic macromolecules that can self-replicate with deadly consequences are prions. Prions are misfolded comparatively small proteins that can cause fatal diseases such as the Creutzfeldt-Jakob Syndrome. Metabolism – chemical transformation – is an essential and distinctive mark for life processes; but there are also numerous inanimate systems in which chemical transformation occurs. Even "evolution through mutation" can occur in the inanimate world. An example is the weathering of rock or, on the molecular basis, the transformation of a meta-stable phase of a chemical species such as ρ-sulfur, ρ-S_6, present in some sulfur bacteria. When ρ-S_6 is deprived of the surroundings providing the metastable situation, i.e., removed from the bacterium, it transforms into a modification that is stable at ambient conditions, namely rhombic α-sulfur, α-S_8.

The following criteria may be coined as paradigms for Life, provided that they are jointly fulfilled in an organizational unit (a cellular system):

a) *Metabolism*, i.e., the formation, transformation and degradation of chemical species via redox and non-redox paths. This includes the self-assembly of molecular parts into functional entities. Directed synthesis, for example with respect to the preferential formation of just *one* specific optical isomer of an amino acid or a sugar, is a noteworthy issue in this context[18]. The utilization of nutrients, as well as the internal alteration of an organism's inventory, ascertains the rebuild and provision of "adapted" molecules in the buildup of tissue components, in the energy balance for the organism, and the breakdown and removal of unnecessary and wintry reaction products,

---

[18] The preferential synthesis of a specific optical isomer is also found in the *inanimate* world. We will come back to this in section 6.1.

including those that are potentially toxic. Metabolic paths are coupled to the production of energy (exothermic processes) and consumption of energy (endothermic processes). The net balance always is energy consumption[19]; metabolism thus depends on external energy sources, such as radiation (light) and thermal energy. In addition, the overall outcome of cellular metabolic processes is connected to a local decrease in entropy (a decrease of disorder, hence an increase of order at the expense of the organism's environment), where "local" refers to an individual organism and its immediate environment, or to a collective, such as a vespiary or bacterial mats. In any case, *non-equilibrium thermodynamics is a signature of life.*

b) Linked to metabolism is *autopoesis,* an internally directed processing network of production, transformation, arrangement and rearrangement of cellular components within the cellular system and (as far as eukarya are concerned) subsystems such as mitochondria, chloroplasts and nuclei, plus cellular compartments, for example lysosomes.

c) *Reproduction,* or self-replication, is an individual's ability to generate copies (offsprings) of itself, coupled to the replication of genetic material as encoded in the parents' information pools of DNA and RNA. The heritage information transferred from the parent to its offspring in the course of reproduction remains essentially unaltered; see, however, (d). Yet, the ability of a living organism to generate a copy (or copies) of itself does not necessarily apply to an *individual* organism within a population. An individual, or even specific groups of individuals, can be infertile, or "exempted" from reproduction in order to fulfill other tasks important for the survival of the complete population. Worker ants and bees are examples of this specific situation.

d) *Evolution:* In the wake of an adaption to novel environmental issues, evolution is an additional indispensable criterion for life. Evolution goes along with (commonly) minor changes – mutations – in the gene and information pools represented by DNA and RNA. These changes

---

[19] Or, to be more precise, consumption of *free* energy, where the free energy change ΔG is related to the change of enthalpy ΔH and entropy ΔS through the Gibbs equation $\Delta G = \Delta H - T\Delta S$ (T is the temperature). The Gibbs free energy ΔG is the *available* energy. ΔG and ΔH are negative by definition when released. Entropy changes ΔS roughly are changes in order. Increasing disorder relates to increasing entropy.

can occur accidentally and thus undirected, but also as a consequence of the necessity to optimize survival and/or to adapt to novel environmental conditions. In either case, the mutation can provide advantages as well as disadvantages with respect to the archetype, with disadvantageous mutations commonly deleted by selection in the struggle for survival. But even in an unchanging environment, multiple mutations can come about, implicating diversion within a species, sometimes within a comparatively short period of time. Respective investigations have recently been undertaken with the intestinal bacterium *Escherichia coli* (Blount 2012). Evolution additionally happens by horizontal gene transfer, i.e., by direct transfer of exogenous DNA between cellular organisms belonging to different species and even different genera and kingdoms, also referred to as transkingdom-sex.

e) *Feedback* is an additional prerequisite for life and survival. Feedback can occur between the cellular organism and environmental conditions (the chemical and energetic status of the organism's surroundings), known as *cognition*, and feedback between individuals belonging to the same population, termed *communication*. Cognition and communication are transmitted via a partially permeable cellular membrane that also serves as a protective barrier. These feedback mechanisms can secure survival, through the adaption to varying environmental conditions by generating genetic diversity within a population. An example is the marine cyanobacterium *Prochlorococcus*, contributing substantially to global photosynthesis. *Prochlorococcus* is composed of hundreds of subpopulations, each with a distinct genomic feature (Kashtan 2014).

How do viruses fit into this picture? The genetic pool of viruses is commonly rather simple, consisting of a single RNA or DNA. Viruses cannot self-replicate independently; reproduction can only be achieved with the help of a host (and often at the expense of the host), such as a bacterium, a protozoa, or cellular constituents in more complex organisms. Also, viruses do not have available their own independent metabolism. Rather, they rely on energy and material sources provided via the host's metabolism, and have therefore been considered to have developed at a later stage in evolution. For evolvement (mutation), viruses commonly remove genes from cellular organisms. Based on these facts, the International Committee on Taxonomy of

Viruses declared in 2000 that viruses are not to be considered alive. Consequently, viruses are commonly subsumed under the term "particles".

Recent evidence might point toward a different direction: Giant DNA viruses, so-called megaviridae such as *Megavirus chilensis*, *Mimivirus*, *Pandoravirus* (Philippe 2013; Figure 5.7) and *Pithovirus* (Legendre 2014) can encode up to 2,500 proteins, and hence more than many bacteria. The DNA of these viruses is either rich in adenine and thymine (*M. chilensis* and *Mimivirus*), or in guanine and cytosine (*Pandora-* and *Pithovirus*). In any case, replication of the viruses depends on the involvement of the nucleus of the host cell. The overall buildup of these viral particles can be comparatively complex. *Pithovirus* for example, 1.5 μm in length and 0.5 μm in breadth, is surrounded by a 60 nm thick envelope with an opening at one end, and a lipid membrane enclosing the internal compartment.

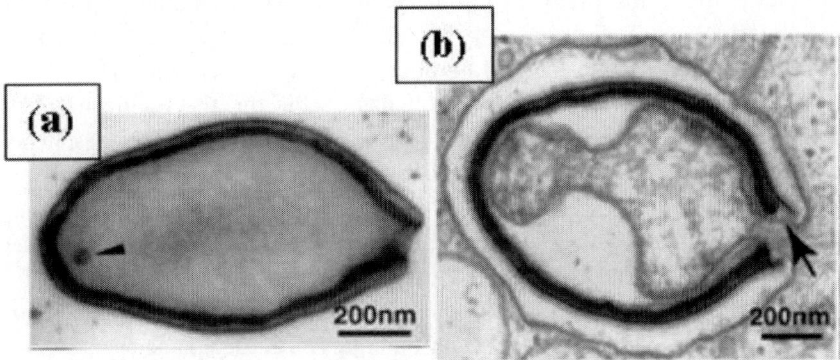

Figure 5.7. (**a**) Electron microscopy image of *Pandoravirus salinus*, showing the shape (a Pandora's box) of the virus, its three-layered envelope plus opening (to the right), and an electron-dense spherical area (arrowhead). (**b**) *P. salinus* internalized in the host vacuole and fused (arrow) with the vacuole membrane (Philippe 2013); © 2013 American Association for the Advancement of Science, reprinted with permission.

Taking into account these recent findings on giant and partially differentiated viruses, megaviridae have provisionally been included as a fourth branch – along with bacteria, archaea and eukarya – in the Tree of Life, Figure 5.1.

5.3. PUTATIVE NON-AQUEOUS AND NON-CARBON LIFE FORMS

So far, we have defined life in a well-acquainted context: Living organisms depend on organic compounds both for the support of life functions and the construction of molecules that provide the basis for tissues and cellular ingredients, and hence the structuring and the operating mode of an organism. Furthermore, we have assumed that these organic molecules are based on organic carbon compounds. Carbon is extraordinary insofar as no other element is involved in the truly gigantic number of molecules that support structure and function in life. The presence of organic carbon compounds – including optically active amino acids in meteorites (for details see section 6.1) – appears to be further reassurance that other worlds employ carbon-based molecules in life forms, provided, of course, that life exists on other worlds.

Similarly, we tend to take it for granted that water is an essential constituent in life, despite the fact that hydrophobic ("water-fearing") effects unequivocally are involved in, or even indispensable for, the stabilization of complex organic architectures such as protein folds and membranes. The presence of water on Mars, likely in the form of both deposits of free water/brines and bound water (water of crystallization), and findings of water also where original anticipation had ruled out its existence – Mercury, the dwarf planet Ceres, the Moon, and Earth's depths – have contributed to expectations according to which the omnipresence of this "elixir" enables, or at least facilitates, life. Water is in fact omnipresent in our planetary system; examples of celestial bodies that carry water, in addition to those just mentioned, are the comets, the icy chunks of Saturn's rings, subsurface oceans on Saturn's moon Enceladus and Jupiter's moon Europa, and the Kuiper-belt dwarf planets (section 2.1). The discovery of water in interstellar clouds, in the protoplanetary discs of T-Tauri stars, and even in AGB carbon stars[20] (where all of the oxygen should be bound to carbon), as well as direct and indirect evidence of water associated with exoplanets (cf. section 7.2) has spurred expectations of finding life forms similar to terrestrial ones elsewhere in the Universe and dwindled anticipations of alternative life forms here and elsewhere.

In the following, we will nonetheless fathom the feasibility of life that does not depend on carbon and/or water. Table 5.1 provides an overview of

---

[20] See the glossary "Types of Stars" for AGB stars and T-Tauri stars.

characteristics of water and various other liquids that might sustain life. I will first summarize, in paragraphs (i) to (v), intrinsic features of water that intimately connect specific properties of water to life functions, and thereafter address other potentially life-supporting liquids.

Table 5.1. Physical properties of water and alternative liquids that might support life

Liquid	Melting/boiling point and temp. range Δ; °C at 10^5 kPa	Heat capacity[a]; J mol^{-1} K^{-1} [b]	Density (liquid state); g cm^{-3} [b]	Dipole moment μ; D
H_2O Water	0/+100 $\Delta = 100$	75.3 (+5°)	1 (+4°)	1.85
H_2S Hydrogen sulfide	-85.7/-60.2 $\Delta = 25.5$	36.4 (+5°)	0.914 (-60°)	0.97
NH_3 Ammomia	-77.7/-33.3 $\Delta = 44.4$	80.0 (+5°)	0.828 (+25°)	1.42
$(HF)_n$[c] Hydrogen fluoride	-83.4/+19.5 $\Delta = 103$	48.7 (0°)	0.99 (+26°)	1.86
HCN Hydrogen cyanide	-13/+26 $\Delta = 39$	70.8 (+17°)	0.687 (+20°)	2.98
CH_3OH Methanol	-97.6/+64.7 $\Delta = 162$	40.9 (+5°)	0.79 (+20°)	1.69
CH_4 Methane	-182.5/-181.5 $\Delta = 21$	53 (-180°)	0.423 (-162)	0
C_2H_6 Ethane	-182.8/-88.5 $\Delta = 94.3$	68.5 (-179°)	0.545 (-89°)	0
CO_2 Carbon dioxide	-78.5[d]/-56.6[e] $\Delta = 22$	109 (0°)	0.77 (20°; 5.6 MPa)	0[f]

[a]The heat that is necessary to change the temperature of one mole of a substance by 1 degree, and hence a measure for heat (energy) that can be stored/released by the liquid. [b]Temperature (in °C) in parentheses. [c]At the boiling point of $(HF)_n$, n is 1 and 6; in the liquid phase, HF has a polymeric structure. The data provided refer to $n = 1$. [d]Sublimation point. [e]Boiling point at 5.2 bar. [f]The two partial dipole moments $^{\delta-}O{=}C^{\delta+}$ of the linear molecule O=C=O cancel each other.

(i) Water (Figure 5.8) is present in its *liquid* state of aggregation over a particularly broad and ambient temperature range, namely = 100 °C under normal conditions (pure water, 10^5 kPa), and hence in a range at which organic molecules are stable *per se*, or can be confined in such a way that they become protected against degradation by heat. This range can be extended to ca. +120 °C under elevated pressure (e.g., in crevices deep inside Earth's crust) and to ca. -15 °C for increased concentrations of ionic substances such as salt (sodium chloride), soda (sodium carbonate), potassium carbonate, and sodium sulfate.

For a still more pronounced depression of the freezing point of water, see (iv). Hydrogen fluoride $(HF)_n$ and the carbohydrate ethane C_2H_6 cover a similar temperature range, at low temperatures, though, where metabolism is dramatically slowed down, if not halted. In contrast, methanol is a competitor for water to be taken into account. Increasing the pressure helps liquefying gases such as CO_2 even at temperatures above 0 °C, an important issue in the context of deep sea CO_2 lakes; *vide infra*.

(ii) The existence of water in the liquid state over a range of a hundred degrees is a consequence of the formation of hydrogen bonds between the water molecules (Figure 5.8, right), resulting in short-lived and steadily rearranged aggregates, or water clusters, the degradation of which requires energy. This normative statement also applies to liquid methanol and, to some extent, to hydrogen fluoride.

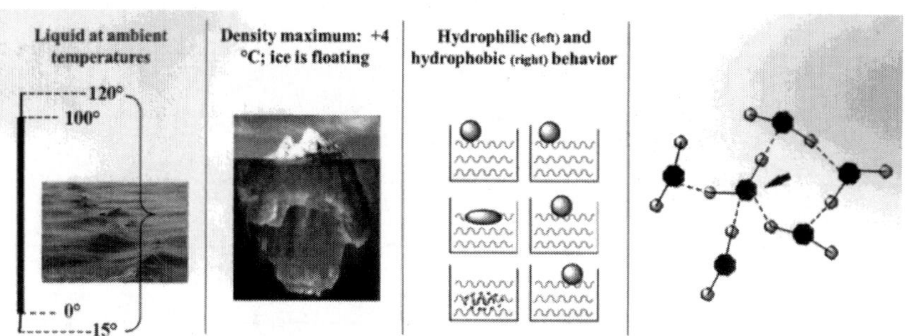

Figure 5.8. Left: Three exceptional properties of water which are supportive to life. See text for additional details. Right: hydrogen-bonding network in water (cutout), providing a tetrahedral environment for the water-oxygens (arrow). The life span of these hydrogen bonds is in the range of picoseconds, i.e., there is permanent rearrangement in the cluster structure.

(iii) Water is one of the rare substances for which the density maximum is associated with the *liquid* state. Water of 4 °C has the highest density, namely 1 g mL^{-1}. As water gets colder and finally freezes at 0 °C, the density decreases to 0.917 g mL^{-1}. Ice is thus 8.3% less dense than liquid water, and hence is floating on water, preventing the freezing-over of a water reservoir of sufficient depth, and thus securing the survival of water-bound living organisms.

(iv) The hydrogen bonding interactions observed between water molecules in the liquid state of water are also effective between water and other polar molecules, resulting in the dissolution (the homogenous dispersal) of substances such as sugar and ethanol. Since water is a dipole, it can further interact with ionic substances, for example sodium chloride NaCl, Eq. (5.1): the association of water molecules around ions is referred to as hydration. Substances that dissolve in water are termed hydrophilic, while those that do not dissolve can either be hydrophobic (oil is an example), or sufficiently stable in the solid (crystalline) state to withstand the breakup by water. In this latter case – examples are insoluble or hardly soluble ionic substances such as scale (calcium carbonate $CaCO_3$) – the energy gain mediated by hydration does not suffice to overcome the strong inter-ionic forces between cations and anions in the crystal lattice.

$$Na^+Cl^- + (x+y)H_2O \rightarrow [Na(OH_2)_x]^+ + [Cl(H_2O)_y]^- \quad x, y \approx 6\text{-}10 \qquad (5.1)$$

The high order within liquid water as represented by the formation of short-lived clusters through hydrogen bonding is disrupted by polar molecules (such as sugar) and by ions provided as an ionic solute (such as sodium chloride NaCl) dissociates into ions (Na^+ and Cl^-). As a result, the freezing point of water can be depressed dramatically. Alkaline metal sulfates and perchlorates, for example, can decrease the freezing point to about -70 °C, allowing water to remain liquid even at the frosty sites of Mars.

(v) An additional intrinsic property of water (and other inorganic polar solvents) is self-dissociation, as indicated by the equilibrium Eq. (5.2a). With respect to life, this property is of some relevance in a variety of physiologically important processes, since water can thus support the regulation of pH values insofar as it can act as a proton donor (or, to be more precise, a donor of hydronium ions H_3O^+; forward reaction in Eq. (5.2a)) as well as a proton acceptor, Eq. (5.2b). The extent of *self*-dissociation of water actually is low: at ambient conditions the concentrations of H_3O^+ and OH^- in the equilibrium (5.2a) are 10^{-7} mol L^{-1} each. Pure water thus is neutral (pH = 7).

$$2H_2O \leftrightarrows H_3O^+ + OH^- \qquad (5.2a)$$

$$H_2O + H^+ \leftrightarrows H_3O^+ \qquad (5.2b)$$

The non-aqueous media hydrogen sulfide H_2S and ammonia NH_3 are less likely suited as habitats for life since, at ambient pressure, they are only liquid at rather low temperatures, possibly too low to render possible life processes. In contrast, liquid hydrogen fluoride HF, hydrogen cyanide HCN and methanol CH_3OH are within the temperature realm where physiological processes proceed sufficiently fast to allow for the sustenance of life. Among these three liquids, methanol is the most "attractive" alternative to water, given the broad range (-98 to +65 °C) where liquid methanol does exist, and a dipole moment comparable to that of water.

The low temperature argument, along with the absence of a dipole moment, should also disqualify methane CH_4, ethane C_2H_6 and carbon dioxide CO_2. CH_4 and C_2H_6 have been included in the overview in Table 5.1 because lakes comprising these liquefied gases have been detected on Saturn's moon Titan (section 7.3), initiating speculations with respect to alternative life forms. The discovery of CO_2 lakes in the Okinawa Trough (Inagaki 2006) has revived discussions with respect to liquid CO_2 as a medium for a microbial habitat: at elevated pressure, CO_2 is liquid at temperatures above zero. The CO_2 lakes at a depth of 1400 m (corresponding to a pressure of 14 MPa) are overlaid by a layer of CO_2-hydrate which again is superimposed and thus kept in place by a pavement consisting of sulfur, quartz and montmorillonite (an aluminum-magnesium silicate). Admixed to the CO_2 are about 10% of H_2S and CH_4. Bacteria and archaea flourish in the intermediate range composed of the CO_2-hydrate and the liquid hydrate-free CO_2, and even – though to a lesser extent – beyond this interface in the hydrate-free zone. Both CO_2 and CH_4 can form cage structures with water molecules, termed hydrates or hydrate clathrates (Figure 5.9). The occurrence of methane ice is well documented for cold high-pressure deep-sea areas on our planet.

Next let us consider the possibility of replacing carbon. The vast majority of molecules that play a role in life is based on carbon. This includes the few inorganic carbon compounds CO, CO_2 and H_2CO_3/HCO_3^-, CS_2, and HCN/CN^-: Carbon dioxide, carbonic acid and hydrogen carbonate are intimately involved in the formation and breakdown of organics such as sugars in the frame of assimilation and dissimilation, respectively. The archaea *Acidianus*, thriving in volcanic sulfataras, utilizes carbon disulfide as a source for generating energy. Carbon monoxide and cyanide are important signaling molecules (apart from being highly toxic at non-physiological concentrations). Sub-molecular and molecular fragments of simple and complex organic molecules are exclusively based on carbon. What are the intrinsic properties that make carbon such an indispensable element in the animate world?

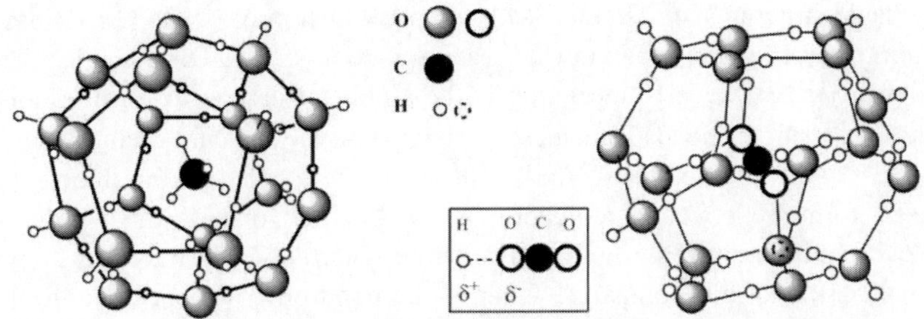

Figure 5.9. The hydrate clathrates of methane (left) and carbon dioxide (right). The carbons of CH_4 and CO_2, in the center of the icosahedral cluster, are represented by a black sphere. Hydrogen bonding between the protons of the water molecules and the two oxygens of CO_2 is facilitated by electrostatic interaction between the partial charges δ^- on the oxygens and the δ^+ on the hydrogens (framed). Respective interactions between the hydrogens of CH_4 and the water oxygens are negligible due to the minor difference in the electronegativities of C and H (and hence negligible polarization in the C–H bond).

Table 5.2. Average bond energies in kJ mol^{-1}. For double and triple bonds, the energy per bond is indicated: the *overall* bond energy for the carbon-carbon triple bond is thus 3×(-280) kJ mol^{-1}, etc.

Carbon-hydrogen and carbon-carbon bonds	Carbon-nitrogen and carbon-phosphorus bonds	Carbon-oxygen and carbon-sulfur bonds
C–H C–C C=C C≡C -413 -348 -307 -280	C–N C=N C≡N C–P -305 -308 -297 -264	C–O C=O C–S C=S -358 -391 -272 -268
Si-Si bonds and bonds of silicon to hydrogen and carbon	Bonds between silicon and nitrogen, oxygen or sulfur	Bonds involving phosphorus
Si–H Si–Si Si=Si Si–C -380 -222 -125 -318	Si–N Si–O Si–S -355 -460 -293	P–H P–P P–C P–O -322 -201 -264 -335

Carbon is the fourth most abundant element in the universe. On our planet, carbon is *the* element that supports the complex chemistry indispensable for the genesis and evolution of life. Carbon atoms readily form bonds to other carbons as well as to atoms such as hydrogen, nitrogen, oxygen, sulfur and phosphorus. All of these bonds are comparable in energy (see Table 5.2).

Carbon-carbon bond formation can generate almost infinitely long chains (examples are polyalkanes and -alkenes) and cyclic structures (such as cyclohexane and benzene) as well as two- and three-dimensional networks, realized for example in polyaromatic compounds, in graphite and diamond.

Matchable structures are featured with the participation of N and O, and have been introduced in the context of fatty acids, carbohydrates, DNA and RNA in section 5.1.1. Particularly intriguing is the ability of carbon to form – along with C–C single bonds (alkanes) – comparatively stable double bonds (alkenes) and triple bonds (alkynes). The molecule C_2, detected in interstellar clouds, formally contains a quadruple bond. Here, one of the bonds is, however, antibonding; the net bond order in dicarbon C_2 is thus just 2 (see sidebar 5.3), and thus compares, as far as the bond order is concerned, to the bond in ethene $H_2C=CH_2$.

To appreciate these specific properties of carbon, and to appraise its possibly unique role in the living environment, it is sensible to compare carbon with silicon and phosphorus, two of its closest neighbors in the Periodic Table (Figure 5.6). Silicon is related to carbon insofar as Si is a member of the same group (group 14), located just underneath carbon, while phosphorus, a group 15 element, is germane to carbon as a consequence of the "diagonal relationship"[21], an affinity that is reflected in the similar electronegativities, namely 2.5 for carbon and 2.1 for phosphorus. With an electronegativity of 1.7, silicon is clearly less electronegative than carbon, and the Si-C bond is thus polarized, meaning that, descriptively speaking, the bonding electron pair is dislocated in the direction of carbon, providing a partial negative charge (δ^-) on C and a partial positive charge (δ^+) on Si.

Since silicon atoms are clearly larger than carbon atoms – the atomic radii are respectively 118 and 77 pm – silicon-silicon bonds (-222 kJ mol^{-1}) are substantially weaker than carbon-carbon bonds (-348 kJ mol^{-1}), rendering long silicon chains energetically less favorable than long carbon chains. Nonetheless, polysilanes (the silicon analogs to alkanes; silane by itself is SiH_4) with up to about 30 silicon atoms in a chain have been prepared in laboratories. The energetic situation becomes more dramatic as double bonds are involved. While the first bond between two atoms comes about by an overlap of two orbitals positioned in the connecting line of these atoms (a so-called σ bond; cf. sidebar 5.3), the second bond (a π bond) is formed between two p-type orbitals perpendicular to the atomic connecting line. This overlap is less favorable; as a result, a π bond experiences a more pronounced decrease in strength than a σ bond as the distance between the atoms increases.

---

[21] Diagonally adjacent elements in the Periodic Table can have similar atomic radii and are then chemically related. This diagonal relationship is particularly distinct for the early row 2 and 3 elements; examples are the pairs Li/Mg and Be/Al.

Organosilicon compounds with a double or even a triple bond, analogous to alkenes and alkynes in carbon chemistry, are thus only stable when protected against consecutive reactions by spacious substituents on Si.

*****

Sidebar 5.3. Orbital description of chemical bonding

In the orbital description of bonding, *s*-type atomic orbitals are spherical-symmetric, *p*-type orbitals dumbbell-shaped. There are three *p* orbitals, directed along the axes of a Cartesian coordinate system, with a nodal point – change from 'plus' (filled lobes) to 'minus' (open lobes) – in the origin of the coordinate system. Hydrogen atoms have only one orbital available (an *s*-type orbital), while carbon atoms possess three *p*-type orbitals in addition to the *s* orbital. The *s* and the three *p* orbitals can "hybridize" to form four sp^3 orbital, with the lobes (only positive lobes are shown) directed towards the corners of a tetrahedron.

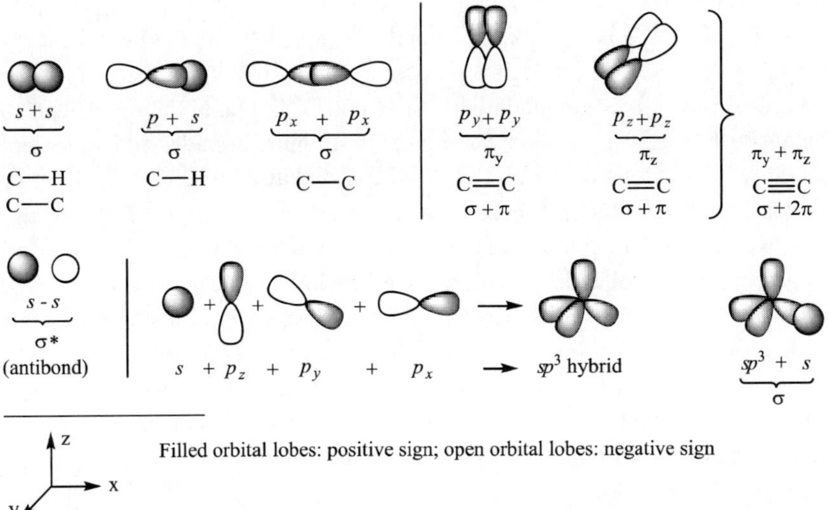

Each orbital of the atoms H and C in their reactive state accommodates one electron. A bond forms by the *overlap* of two orbitals of two elements, H and C in this specific case, to provide a molecular orbital, occupied by two electrons, and representing a single bond symbolized by a valence dash. C and H can form a single bond only, coming about by the overlap of either two *s*

orbitals, or a hydrogen-*s* and a carbon-*p* orbital oriented in the *x* direction (p_x), or the *s* orbital of H and an sp^3 hybrid of C. In any case, the resulting bond is a σ bond, with the maximum electron density in the area connecting the two bonding partners. Overlap of two carbon *p* orbitals directed along the *y* or the *z* axis (and hence perpendicular to the atomic connecting line) provides a π bond, with the maximum electron density below and above the line directly connecting the two carbons (the σ bond). Carbon atoms can thus be linked by a single bond (σ) only, a double bond (σ + π) or a triple bond (σ + 2π). In addition to the '*s* plus *s*' σ bond, an '*s* minus *s*' anti-bond σ* can come in. This is the case in the interstellar molecule dicarbon C_2, where there are two π bonds, a σ bond and a σ* bond, the latter two annulling each other and giving rise to a net double bond in C_2, exclusively of π character, and thus different from the σ + π double bond in ethene $H_2C=CH_2$.

***

Si-H, Si-C, Si-N and Si-S bond energies do not differ to a great extent from the respective bond energies for carbon. The bond energy for the Si-O bond (-460 kJ mol^{-1}) is, however, distinctly greater than the C-O bond energy (-358 kJ mol^{-1}). Silicon thus has a particularly high affinity to oxygen. In the presence of oxygen, compounds with Si-H and Si-C bonds therefore tend to spontaneously convert to siloxanes or even silicium dioxide. Siloxanes are essentially inert compounds with (R)Si-O-Si(R) linkages, where "R" stands for a carbon-based substituent. Silicon dioxide, or silica $(SiO_2)_\infty$, is an essentially unreactive polymer, in contrast to the comparatively reactive mononuclear and thus gaseous CO_2 molecule. Sea-water, with its high contents of sodium ions, can contain up to 0.1 mM of "dissolved SiO_2" – actually silicic acid H_4SiO_4 and hydrogen silicate $H_3SiO_4^-$. Diatoms exploit this dissolved silica to manufacture protective shells of the approximate composition $SiO_n(OH)_{4-2n}$ (kieselgur) embedded in a protein matrix. As mentioned in sidebar 5.2, silicate is also a constituent of other scaffolds, and involved in the early stages of bone formation.

Our next candidate for a *carbon analog* in the animate world is phosphorus. The atomic radius of phosphorus, 111 pm, compares to that of silicon (118 pm), i.e., as far as the stability of the P-P bond is concerned, there are similar concerns to those discussed for the Si-Si bonds. Phosphanes are in fact quite reactive, and double and triple bonds between P atoms are disfavored. Along with linear phosphanes P_nH_{2+n} (n = 2-6) $\equiv H_2P(PH)_mPH_2$

(where m = 0-4), cyclic phosphanes such as P_6H_6, and phosphanes with cage structures are known. An example of the latter is P_9H_5. Sufficiently more stable than phosphanes are phosphazenes $(Z_2P=N)_n$, where Z can be OR or NR_2. Phosphazenes can be cyclic (n = 3) or linear (n → ∞). R is an organic residue such as methyl or ethyl; phosphazenes are thus not inorganics in a strict sense.

We have seen that phosphate as an activator of numerous biologically relevant molecules and molecular assemblies. A potential substitute for phosphate is vanadate. The anions vanadate and phosphate are built almost analogously, insofar as they are both tetrahedral and of almost the same size. There is, however, a difference in the protonation state and thus the charge of the anions in neutral (pH 7) aqueous solutions: While vanadate is present almost exclusively as $H_2VO_4^-$, phosphate comes as an approximately equal mix of $H_2PO_4^-$ and HPO_4^{2-}. The similarity between vanadate and phosphate implies that vanadate principally can function – at low concentrations in the µM range – as a regulator for phosphate-dependent enzymes such as ATPases (Rehder 2008). In other words: vanadate may well substitute for phosphate. There are, however, also definite differences between vanadate and phosphate: The transition element vanadium can easily switch between the oxidation states +V and +IV (and, to some extent, +III), and extend its coordination sphere to five- and six-fold coordination, while the coordination number 5 is only a transient state for phosphorus in the phosphate anion. In any case, vanadium has been detected, in the form of vanadium oxides VO_x (where x commonly equals 1), in the atmospheres of exoplanets (Rehder 2011).

In conclusion, although silicon and, in particular, phosphorus undoubtedly are involved in processes of life as we know it, their chemistry appears to be insufficiently versatile to qualify for a substitute of carbon in life, and thus to be considered an alternative for carbon in molecules that store and transfer information, or are involved in metabolic paths. On our planet, where silicon comes second in abundance (after oxygen), life has "chosen" carbon rather than silicon as a basis for building life molecules.

The cosmic abundance of carbon (4th most abundant element) exceeds that of silicon – number 8 in the list of cosmic elements – by a factor of approximately seven. The cosmic abundance of phosphorus is only one thousandth that of silicon. Of the approximately 180 molecules that have so far been detected in the interstellar medium, eleven contain silicon (SiH, SiC, SiN, SiO, SiS, SiCN, SiNC, *cyclo*-SiC$_2$, *cyclo*-SiC$_3$, SiC$_4$, SiH$_4$) and are thus potential – though exceedingly unlikely – sources for an animate silicon world. This does not, however, exclude the possibility of silicon having been *involved* in the origin and development of life on Earth, and equally on other planets; provided that there is life at all on other planets. We will come back to this involvement of silicates in the context of "clay organisms" in section 6.1. Phosphorus is represented in the interstellar medium by six molecular compounds: CP, PN, PO, C_2P, HCP and PH_3.

5.4. EXTREMOPHILES

We have pointed out in section 5.2 that life goes along with high order (low entropy), and thus represents the fragile state of thermodynamic and chemical disequilibrium. Living organisms are constantly trying to minimize this disequilibrium situation by dwarfing unfavorable environmental conditions. We thus tend to associate environmental factors that enable life with moderate situations, moderate with respect to temperature, pressure, salinity, pH, radiation, and the presence of toxic metals/metalloids and molecules. We also tend to deny the possibility of blooming life (and, of course, our treasure trove of experience appears to affirm this) as extremes in these six parameters are encountered, as well as with drought and either the absence or the presence of oxygen, depending on whether aerobic or anaerobic conditions prevail. But organisms can adapt, at least to a certain degree, to a hostile environment – "hostile" as opposed to a moderate, hospitable environment. A prominent example is the adaption of a multitude of the present life forms, multicellular life in particular, to oxygen O_2 as an emerging constituent in our atmosphere, in water bodies and in sediments, an adaption that started off with the Great Oxygen Event (or, less euphemistically, the Oxygen Catastrophe), when cyanobacteria "invented" photosynthesis about 2.4-2.1 billion years ago (see also Figure 5.10).

We can define moderate conditions ("moderate" from our subjective point of view) in the following way: (i) Average temperatures around 15 °C, with seasonal and geographical variations of up to ca. +45 (where proteins can start to denature) and down to -2 °C (the freezing point of sea water at sea level); (ii) pH values near pH 7, i.e., no excessive alkalinity or acidity; (iii) pressure in the range of 30 kPa (atmospheric pressure at an altitude of ca. 10 km) to 100 kPa (at sea level) and a hydrostatic pressure of ca. 10^5 kPa for a sea water depth of 1 km; (iv) moderate UV flux, and moderate γ and X-ray radiation (2 mGy a^{-1} at ground level); (v) a non-toxic environment, i.e., low levels of toxic elements such as arsenic, cadmium, lead and mercury; (vi) moderate osmotic strain, i.e., low to medium salinity as provided by the concentration of "salts" in fresh water and sea water. In addition, primary energy sources have to be provided, commonly the light and warmth (IR radiation) delivered by the Sun and/or the interior heat of our planet, including heat released in the course of radioactive decay.

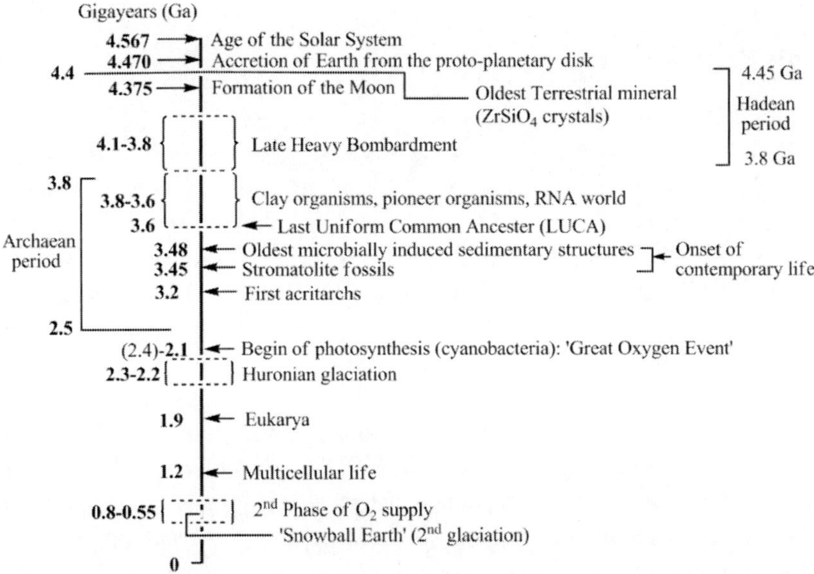

Figure 5.10. A time-line (not to scale) for the development of geological features and life on Earth. Recent new insight has been provided for the age of the Moon (Jacobson 2014), the oldest fossils (Noffke 2013), and the Great Oxygen Event (Kerr 2014). Acritarchs (Buick 2010) represent a somewhat artificially defined group of microfossils, including unicellular algae and dinoflagellates, for which the systematic indexing is poorly known. Whether Earth actually froze from pole to pole ("Snowball Earth") about 0.8-0.6 billion years ago is still being debated (Kerr 2010). The Hadean period (4.45-3.8 Ga) derives from Hades, the Greek god of the underworld.

Finally, basic "feed" molecules need to be around, such as O_2 for aerobic organisms, CO_2 for organisms relying on photosynthesis, or CH_4 for methanogens.

An organism relying on light as the energy source is termed a phototrophic organism, while an organism resorting to the energy delivered in the frame of chemical processes is referred to as chemotroph. Typical examples of chemotrophic processes are the oxidation of hydrogen sulfide to sulfate, ammonium ions to nitrate, and molecular hydrogen to protons; Eqs. (5.3) to (5.5).

$$HS^- + 4H_2O \rightarrow SO_4{}^{2-} + 9H^+ + 8e^- \tag{5.3}$$

$$NH_4{}^+ + 3H_2O \rightarrow NO_3{}^- + 10H^+ + 8e^- \tag{5.4}$$

$$H_2 \rightarrow 2H^+ + 2e^- \tag{5.5}$$

In all three processes shown here, electrons are delivered. Since the electrons are coming from *inorganic* sources, the processes are termed lithotroph: *lithos* for stone or rock (inorganic), and *troph*e for nutrition. Alternatively, an organic electron source can be exploited. An example is the oxidation of glucose to pyruvate, Eq. (5.6). Such reactions are referred to as organotroph. A résumé on the various "*trophs*" is also provided in sidebar 5.3.

$$\tfrac{1}{2}C_6H_{12}O_6 \rightarrow CH_3\text{-}C(O)\text{-}CO_2{}^- + 3H^+ + 2e^- \tag{5.6}$$

Energy and electrons are intimately involved in the buildup and breaking-down of organic matter. As, in the course of the generation of organics, the carbon source is an organic compound, such as acetate, the organism involved in this process is termed a *heterotroph*. Conversely, if the carbon source is inorganic, commonly carbon dioxide or hydrogen carbonate, the respective process is referred to as *autotroph* ("self-feeding"). Photosynthesis – the conversion of CO_2 and water into glucose and oxygen with light as the energy source – is a typical example of autotrophic activity.

A microorganism that uses light as the energy source, hydrogen as the electron source and acetate as the carbon source to synthesize organic molecules thus is a photo-litho-heterotroph. In contrast, the microbial communities detected in the interfacial area of the liquid CO_2-methane phase and the solid CO_2-water phase in the Okinawa Trough briefly addressed in section 5.2 are *chemo*-litho-heterotrophs, i.e., they do not use light, but rather the energy provided by a chemical reaction to synthesize "their" organics. Respective reactions are the oxidation of hydrogen sulfide, Eq. (5.3), or methane, Eq. (5.7). In Eq. (5.7), the oxidant is sulfate; sulfate is reduced to hydrogen sulfide:

$$CH_4 + SO_4{}^{2-} + H^+ \rightarrow CO_2 + HS^- + 2H_2O. \tag{5.7}$$

The energetically rather favorable oxidation of sulfide, Eq. (5.3), is also used by those extremophiles among the prokarya that thrive in brines (halophiles) and in extremely acidic habitats (acidophiles). Bacteria that are comfortable both with brines and with alkaline media (pH values up to 12) are termed haloalkaliphiles. The suffix *phile* indicates that the organisms favor the extreme over normal conditions; for a classification of the various extremophiles, see also sidebar 5.4.

The brines are typically soda lakes, commonly containing, along with soda (Na_2CO_3) and sodium bicarbonate (sodium hydrogen carbonate $NaHCO_3$), as well as sodium chloride $NaCl$ in an overall concentration of up to 5 M or, assuming an approximately equal mix of $NaHCO_3$ and $NaCl$, a salt load of about 35%.

This is also the salt concentration in the Dead Sea (and likely in the subsurface oceans of Titan), and corresponds to a saturated salt solution. The Dead Sea, carrying a mix of $NaCl$, $MgCl_2$, $CaCl_2$ and KCl, is by no means dead; rather, halotolerant algae and halophilic bacteria are at ease. Hypersalinity (very high salt concentrations) exerts osmotic stress; halophiles have developed strategies to protect themselves against the high external salt loads.

***

Sidebar 5.4. Classification of microorganisms

According to trophisms (I), and extreme conditions (extremophiles, II):

I. Trophisms

Chemotroph vs. phototroph: energy to sustain life is delivered by a chemical reaction (chemotroph), or through light (phototroph);

Lithotroph vs. organotroph: reduction equivalents (electrons) are delivered by an inorganic (lithotroph), or by an organic ion/molecule (organotroph);

Heterotroph vs. autotroph: organic molecules are synthesized from an inorganic carbon source (heterotroph), or from an organic carbon source (autotroph).

II. Extremophiles

Acidophiles: The growth optimum is in the highly acidic pH range (commonly pH < 3, down to 0). Typical habitats are sulfataras, deep-sea vents, and mining waters.

Alkaliphiles preferentially thrive at pH values > 9-10 (up to 12), such as are encountered in soda lakes.

Halophiles grow at salt concentrations exceeding the concentration in sea water (i.e., > 0.6 M) up to saturated salt solutions (ca. 5 M, as in the Dead Sea).

Piezophiles (also known as barophiles): They thrive in high pressure regions encountered close to deep sea hydrothermal vents, subglacial lakes and subsurface clefts deep in Earth's crust. They tolerate pressures up to 130 MPa, and possibly even more.

Psychrophiles (also termed cryophiles): These organisms survive temperatures down to at least -10 °C, and commonly do not tolerate temperatures above +20 °C.

Thermophiles prefer, or even thrive exclusively at, temperatures where proteins of organisms living at ambient temperature coagulate. 'Normal' thermophiles have their growth optimum around +50 to +60 °C; *hyper*thermophiles from ca. +80 up to +122 °C.

Xerophiles resist extremely dry environments for sometimes rather long periods of time (up to a century and possibly more).

***

The situation becomes even more extreme as other stress factors add to hypersalinity, such as low temperature, low supply of organics and high contents of ferrous iron. Psychrophiles, also termed cryophiles, can adapt to temperatures down to about -10 °C. High salt concentrations result in a substantial depression of the freezing point of water, an essential requisite to avoid damage of cells caused by the shear forces exerted by water ice. Remember that as water freezes, the volume per mass *increases* (a *decrease* in density). A subglacial, anoxic and organic-starved pool of marine brine (1.375 M in chloride) at a depth of ca. 4 km, a temperature of -5 °C, and a rather high concentration of Fe^{2+} (3.4 mM) has been shown to harbor an active microbial consortium, thriving and prospering in isolation for at least 1.5 million years (Milucki 2009). Environmental conditions like these[22] are kept "warm" by geothermal energy, and may serve as a model for subglacial lakes on, for example, Jupiter's moon Europa and Saturn's moon Enceladus. Microbial life

---

[22] See also the Antarctic Lake Vostok, presently under investigation with respect to, *inter alia*, its potential microbial community.

also persists in ice-covered lakes and desiccated soil in Antarctic dry valleys with air temperatures going down to -40 °C in the winter (Dartnell 2010). In addition, the microorganisms living in these arctic deserts survive the high UV flux in periods of temporary ozone depletion.

Noteworthy in the context of survival strategies despite extreme radiation is *Deinococcus radiodurans*. This terrestrial bacterium accepts a γ ray exposure 5,000 times greater than the dose tolerated by any other terrestrial organism. Since this level of radiation had never existed on Earth, *D. radiodurans* has nourished speculations in view of its potential extraterrestrial origin (Ortiz 2011). However, the γ ray resistance may also be a side-effect of a resistance developed against long periods of drought: DNA damage induced by drought resembles that which is due to radiation, and similar mechanisms may have developed for protection against aridity and highly energetic electromagnetic radiation.

With respect to potentially freezing temperatures and high radiation, this extreme situation is reminiscent of what microbial life might encounter on Mars, with its surface and subsurface temperature close to or even below the freezing point of water, high exposure to UV and γ radiation as a consequence of the particularly thin Martian atmosphere, and high salt loads, including salts that exert oxidative stress, chlorates in particular. Survival experiments with extremophiles simulating, to some extent, Martian surface conditions have been carried out at the Mars Desert Research Station in Utah/USA (Direito 2011). These investigations have been accomplished with xerophilic, halophilic, kryophilic and acido/alkaliphilic bacteria, fungi, algae and protozoa.

Microbial communities have also been detected in deeply buried marine sediments in the North Pacific Gyre (Røy 2012). These aerobic microbes have to cope with an extremely poor supply of organics and oxygen, and thus have exceptionally low metabolic and reproduction rates, resulting in a biomass turnover of only once every few hundred years at the best. But even multicellular organisms can dwell under extreme conditions and in hypoxic settings. As an example, nematoda reside in fracture water in South African mines in depths down to 3.6 km (Borgonie 2011), where the oxygen pressure is as low as 2.9 kPa, corresponding to an atmospheric pressure at an altitude of about 10 km.

Intra-terrestrial, pressure-loving microorganisms (termed baryophiles or piezophiles) have been retrieved from basaltic and sedimentary rocks at depth down to 10 km, where the pressure goes up to 80 MPa. Contemporaneously, these microbes are thermophiles – they can tolerate temperatures up to +120

°C. Basaltic rock is particularly devoid of organics. The microorganisms are thus lithoautotrophs, i.e., they are reliant on carbonate as the carbon source, and they exploit the oxidation of hydrogen, Eq. (5.5), as an energy source. A likely source of hydrogen is the redox reaction between ferrous iron and water, Eq. (5.8).

$$2Fe^{2+} + 2H_2O \rightarrow 2FeO(OH) + H_2 \tag{5.8}$$

Alikaliphiles and acidophiles can tolerate pH values up to pH $\approx$ 11-12 (alkaliphiles) and down to pH $\approx$ 0 (acidophiles), corresponding to the pH of a 1 mM aqueous sodium hydroxide solution and a 1 M aqueous HCl solution, respectively. The soda lakes mentioned above are habitats for alkaliphiles. The gastric mucosa of most humans is colonized by *Helicobacter pylori*, an acidophile that is involved in the regulation of the stomach pH (pH $\approx$ 2), but can occasionally cause ulcers.

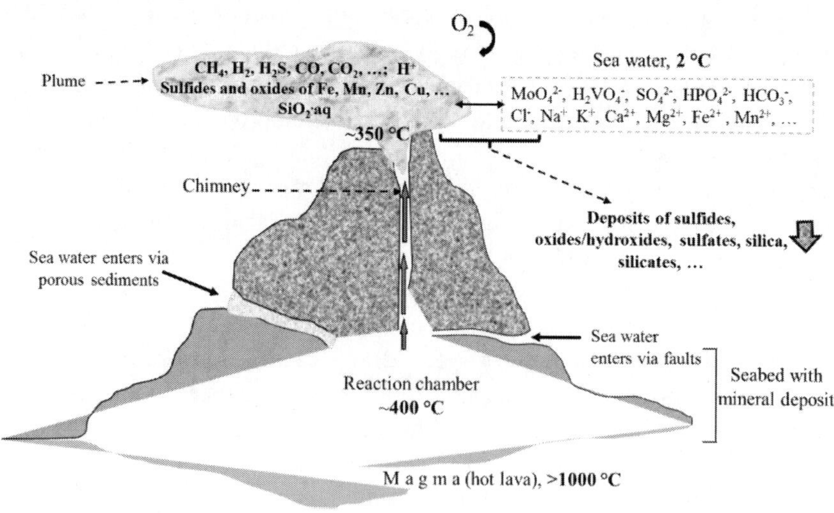

Figure 5.11. Schematic view of a deep-sea hydrothermal vent forming chimney-like deposits (black/white smoker) at a depth of several kilometers. O_2, mainly stemming from photosynthetic processes near the surface, arrives at the reaction zone via diffusion. The thermophilic, lithochemoheterotrophic microbiota thriving in the proximity of the black smokers are adapted to an anoxic environment.

Acidophiles that concomitantly are thermophiles typically dwell in volcanic sulfataras, and in the immediate proximity of deep-sea black and white smokers (see Figure 5.11). These smokers form from hydrothermal vents nourished by volcanic activity just underneath the sea floor surface in deep-sea ocean areas. Sea water entering the hot reaction chambers via faults in the crust of the sea floor or via porous sediments is heated to about 400 °C, conditions where water can react with minerals to form gaseous components and particulate matter.

The formation of gaseous components (such as H_2S, CH_4, CO, CO_2, H_2 and NH_3) drastically increases the internal pressure, leading to a discharge of gas along with 'dust' predominantly consisting of sulfides of the metals Fe, Ni, Mn and Cu, and giving rise to the formation of turrets of blackish deposits, the "black smokers". Alternatively, the ejecta can be dominated by hydrated silica, the sulfates of Ba and Ca, and zinc sulfide, providing white deposits ("white smokers"). Deposits are additionally formed from the interaction of (gaseous) components in the plumes and ingredients from sea-water.

Complete ecosystems flourish in the immediate proximity of these hydrothermal vents. The gases relayed to the surrounding sea water in these scenarios serve as nutritional and energy sources essentially for thermophilic bacteria and archaea which in turn are grazed by eukaryotes such as clams and tube worms. Methanogens, for example, reduce CO_2 to CH_4, while methanotrophs exploit the reverse reaction, i.e., the oxidation of CH_4 to CO_2. The delivery of H_2S and ferrous sulfide FeS in the frame of the activity of the black smokers is of particular interest also in the context of scenarios for the initiation of Life (see the Wächterhäuser world, which will be covered in some detail in section 6.3). Record holders among the thermophiles, which might be LUCA's closest relatives, are the archaea *Methanopyrus kandleri* and *Geogemma barossi*. *M. kandleri* is a methanogen that reproduces at 122 °C and a pressure of 40 MPa, corresponding to a water depth of 4 km. Reducing agent can be H_2; the respective gross reaction is depicted in Eq. (5.9). *G. barossi* reproduces at temperatures of up to 121 °C; it metabolizes by reducing oxidic ferric (Fe^{3+}) compounds to ferrous (Fe^{2+}) iron, Eq. (5.10).

$$CO_2 + 4H_2 \rightarrow CH_4 + 2H_2O \tag{5.9}$$

$$FeO(OH) + 1\tfrac{1}{2}H_2 \rightarrow Fe^{2+} + 2H_2O \tag{5.10}$$

Contrasting thermophiles and psychrophiles, where the organism *as a whole* has to adapt to the extreme situation, other extremes such as excessive salt loads, high/low pH, and pronounced concentrations of toxic metal ions are coped with by the organisms through the regulation of inlet/outlet mechanisms that reside in the organism's protective *membrane*. Similarly, the cellular membrane protects an organism against long periods of drought and exposure to extreme doses of UV and γ radiation.

Geochemical reactions taking place in hydrothermal vents with their ion gradients provide energy in a manner comparable to the energy release (and intermittent storage of this energy in the form of adenosine triphosphate) when ions such as H^+ and Na^+ are pumped across a cellular membrane. This energy release has spurred speculations with respect to an origin of life inside the range of these vents. Energy production is particularly efficient as serpentinization[23] comes in (Martin 2014). The net reaction in the frame of serpentinization is the oxidation of ferrous to ferric iron by seawater, Eq. (5.11). The energy released in this process is coupled to the reduction of carbon dioxide to methane (by methanogens) or formate/acetate (by acetogens).

$$Fe^{2+} + H_2O \rightarrow Fe^{3+} + \tfrac{1}{2}H_2 + OH^- \tag{5.11}$$

Likewise, there are microorganisms that not just tolerate but even thrive at high concentrations of metals and metalloids that normally are toxic because they compete with life metals. Examples are bacteria dwelling in effluents of mining areas contaminated with arsenate $HAsO_4^{2-}$, cadmium Cd^{2+}, mercury Hg_2^{2+}/Hg^{2+} and/or lead $Pb^{2+/4+}$. But also in the case of just ambient concentrations of toxic metal ions, organisms sometimes resort to such metal ions. Thus, some marine diatoms utilize cadmium instead of zinc in the active center of carboanhydrase, an enzyme that catalyzes the interconversion between CO_2 and HCO_3^-. Yet other organisms have developed specific protective schemes against toxic metals and metalloids: Bacterial populations in the Mono Lake in California, with a high load of arsenate (about 0.2 mM), prevent the transport of toxic arsenate across the cell membrane by a specifically designed transporter which distinguishes between toxic arsenate and life-sustaining phosphate.

Finally, it should be noted that strategies for the survival at extreme conditions are not necessarily restricted to microbes. An example is the arctic

---

[23] Serpentinization is the conversion of olivine $(Fe,Mg)_2SiO_4$ by water into magnetite Fe_3O_4, silicon dioxide, hydrogen and serpentine. Serpentine is a magnesium silicate.

midge *Belgica antarctica* which tolerates freezing cold, high UV radiation, osmotic stress and transient dehydration. *B. antarctica* has a particularly small genome, although a similar number of genes as related insects, suggesting that the ability of the midge to adapt to these extreme situations commonly hostile to life correlates with constraints in genome *architecture* (Kelley 2014).

SUMMARY

We have an intuitive understanding of the notions "alive" and "not alive", or "animate" and "inanimate" (Bains 2013). Although there is no concise and unequivocal definition of what we should envisage when resorting to the term 'Life', we can coin a few manifestations allowing for differentiation between living organisms on the one hand, and inanimate matter on the other hand. These manifestations include metabolism (the setting and cutback of organic matter) coupled to a net energy gain, reproduction, evolution (adaption to novel environmental issues) via mutation and horizontal gene transfer, and feedback with colony members and the environment. Archaea, bacteria and eukarya fit into this picture of life; as yet, viruses are commonly not considered living organisms.

Life as we know it rests upon organic matter, i.e., carbon-based molecules and molecular assemblies, and upon water as a life-sustaining medium. The main groups of molecular assemblies are proteins (representing the bulk of organic matter), lipids (for energy storage and the demarcation of cells via cellular membranes), carbohydrates (for energy supply and building blocks in, e.g., RNA and DNA), and nucleic acids. Deoxynucleic acids (DNA) store information, while ribonucleic acids (RNA) are responsible for, *inter alia*, information transcription, such as the information stored in DNA into a functional protein. Artificial alternatives of the natural DNA have lately been introduced successfully into strains of the intestinal bacterium *Escherichia coli*. Along with organics, inorganics are required to sustain life. Examples of indispensable inorganics are phosphate, various transition metal ions in the active centers of enzymes, and metal ions such as Na^+, K^+, Ca^{2+} and Mg^{2+} in osmotic balances and energy transfer.

The importance of water as a medium to sustain life is based on its intrinsic properties: the broad and ambient temperature range where water is liquid; its dipole character, allowing for the dissolution of ionic as well as non-ionic hydrophilic molecules; and its density maximum at 4 °C, i.e., at a temperature at which water is still a liquid and solid water (ice) thus floats.

Solutes that potentially are alternatives to water at extreme conditions, encompassing conditions on non-terrestrial bodies, include carbon dioxide, hydrogen fluoride, and alkanes. The unique role of carbon in the buildup of organic matter is based on, among other things, its ability to form stable carbon-carbon chains of almost infinite length, and carbon's ability to coin sterically distinct structural sequences through carbon-carbon double and triple bonds. The two immediate neighbors of carbon in the Periodic Table, silicon and phosphorus, sometimes discussed as alternatives for complex life molecules, are quite restricted in this respect.

Many prokarya on our home planet thrive at extreme conditions that might reflect similar conditions on exoplanets of about the same size and thus comparable composition to Earth, and at a distance to their sun that allows for the presence of liquid water. These so-called extremophiles can tolerate temperatures up to 120 °C and down to -10 °C, pressures up to about a thousand-fold of the atmospheric pressure at ground level, extremely high radiation (UV and γ) exposure, high alkalinity (pH $\approx$ 11) and acidity (pH $\approx$ 0), saturated salt solutions, and a particularly low supply of nutrients. Sulfataras and deep sea vents, the Dead Sea and soda lakes, subglacial lakes and minor clefts deep in Earth's rocky crust are indigenous examples of the habitats of extremophiles.

REFERENCES

Bains, W. (2013): What do we think life is? A simple illustration and its consequences. *Int. J. Astrobiol.* 13:101-111.

Blount, Z.D., J.E. Barrick, C.J. Davidson, and R.E. Lenski (2012): Genomic analysis of a key innovation in an experimental *Escherichia coli* population. *Nature* 489:513-518. See also E. Penniski (2013): The man who bottled evolution. *Science* 342:790-793.

Borgonie, G., A. Garcia-Moyano, D. Litthauer, et al. (2011): Nematoda from the terrestrial deep subsurface of South Africa. *Nature* 474:79-82.

Buick, R. (2010): Ancient acritarchs. *Nature* 463:885-886.

Dartnell, L.R., S.J. Hunter, K.V. Lovell, et al. (2009): Low-temperature ionizing resistance of *Deinococcus radiodurans* and Antarctic dry valley bacteria. *Astrobiology* 10:717-732.

Direito, S.O.L., P. Ehrenfreund, A. Marees, et al. 2011: A wide variety of putative extremophiles and large beta-diversity at the Mars Desert Research Station (Utah). *Int J. Astrobiol.* 10:191-207.

Inagaki, F., M.M.M. Kuypers, U. Tsunogai, et al. (2006): Microbial community in a sediment-hosted CO_2 lake of the southern Okinawa Trough hydrothermal system. *Proc. Natl. Acad. Sci.* 103:14164-14169.

Jacobson, S.A., A. Morbidelli, S.N. Raymond, et al. (2014): Highly siderophile elements in Earth's mantle as a clock for the Moon-forming impact. *Nature* 508:84-87.

Kashtan, N., S.E. Roggensack, S. Rodrigue, et al. (2014): Single-cell genomics reveals hundreds of coexisting subpopulations in wild *Prochlorococcus*. *Science* 344:416-420.

Kelley, J.L., J.PT. Peyton, A.-S. Fiston-Lavier, et al. (2014): Compact genome of the Antarctic midge is likely an adaption to an extreme environment. *Nature Commun.* DOI: 10.1038/ncomms5611.

Kerr, R.A. (2010): Snowball Earth has melted back to a profound wintry mix. *Science* 327:1186.

Kerr, R.A. (2014): New look at ancient mineral could scrap a test for early oxygen. *Science* 343:960.

Legendre, M., J. Bartoli, L. Shmakova, et al. (2014): Thirty-thousand-year-old distant relative of giant icosahedral DNA virus with a pandoravirus morphology. *Proc. Natl. Acad. Sci.* 111:4274-4279.

Li, L., C. Francklyn, C.W. Carter jr. (2013): Aminoacylating urzymes challenge the RNA world hypothethis. *J. Biol. Chem.* 288:26856-26863.

Malishev, D.A., K. Dhami, T. Lavergne, et al. (2014): A semi-synthetic organism with an expanded genetic alphabet. *Nature* 509:385-388.

Martn, W.F., F.L. Sousa, N. Lane (2014): Energy at life's origin. *Science* 344:1092-1093.

Mikucki, J.A., A. Pearson, D.T. Johnston, et al. (2009): A contemporary microbially maintained subglacial ferrous ocean. *Science* 324:397-400.

Nielsen, P.E., and M. Egholm (1999): An introduction to peptide nucleic acid. *Current Issues Molec. Biol.* 1:89-104.

Noffke, N., D. Christian, D. Wacey, et al. (2013): Microbially induced sedimentary structures recording an ancient ecosystem in the ca. 3.48 billion-year-old Dresser Formation, Pilbara, Western Australia. *Astrobiology* 13:1103-1124.

Ortiz, P.S. (2011): Antarctic microbes live life to the extreme. *Nature* doi:10.1038/news.2011.207

Philippe, N., M. Legendre, G. Doutre, et al. (2013): Pandoravirus: Amoeba viruses with genomes up to 2.5 Mb reaching that of parasitic eukaryotes. *Science* 341, 281-286.

Pinheiro. V.B., A.I. Taylor, C. Cozens, et al. (2012): Synthetic genetic polymers capable of heredity and evolution. *Science* 336:341-344.

Plaxco W., and M. Gross (2011): Astrobiology. The John Hopkins University Press, Baltimore, U.S.A.

Rehder, D. (2008): Bioinorganic vanadium chemistry. John Wiley & Sons, Chichester.

Rehder, D. (2011): A possible role for extraterrestrial vanadium in the encounter of life. *Coord. Chem. Rev.* 255:2227-2231.

Rehder, D. (2014): Bioinorganic Chemistry. Oxford University Press, Oxford, U.K.

Røy, H., J. Kallmeyer, R.R. Adhikari, et al. (2012): Aerobic microbial respiration in 86-million-year-old deep-sea red clay. *Science* 336:922-925.

Schöning, K.-U., P. Scolz, S. Guntha, et al. (2000): Chemical etiology of nucleic acid structure: The α-threofuranosyl-(3'$\rightarrow$2') oligonucleotide system. *Science* 290:1347-1351.

Chapter 6

INITIATION AND PROGRESSION OF LIFE

As has been outlined in Chapter 5.1, life is based – from a material-related point of view – on rather complex molecular assemblies: (i) peptides and proteins, formed by linking amino acids; (ii) lipids, generated by condensation between glycerol and usually long-chain fatty acids or alcohols; (iii) polysaccharides such as starch and cellulose based on sugars, glucose in particular; and (iv) polymeric molecules required for information storage and transfer: DNA/RNA, assembled from deoxyribose/ribose, phosphate and just five specific nucleobases. The basic building blocks for life are thus organic molecules, in particular amino acids, glycerol and carbonic acids or alcohols, sugars, and N-heterocycles; plus the inorganic linker molecule and energizer phosphate.

There are (at least) two conceivable scenarios for the supply of the molecular fragments for the assemblage of these modules of life to those planets that dispose of the potentiality to generate and to support life: In one of these scenarios, the building blocks are genuine to the interstellar clouds from which the planetary systems formed; in an alternative scenario, the basic molecular fragments for the assembly of molecules that constitute life are generated from more basic precursor molecules on the planet itself. Several popular (and equally scientifically founded) hypotheses for this second scenario will be outlined in section 6.2. In section 6.1, the potential origin of building blocks for life molecules in the interstellar medium – and thus the first scenario – is considered. In section 6.3, facts and circumstances will briefly be addressed that could have contributed to the evolvement of basic molecules afforded to produce units for more complex components in the genesis of Life.

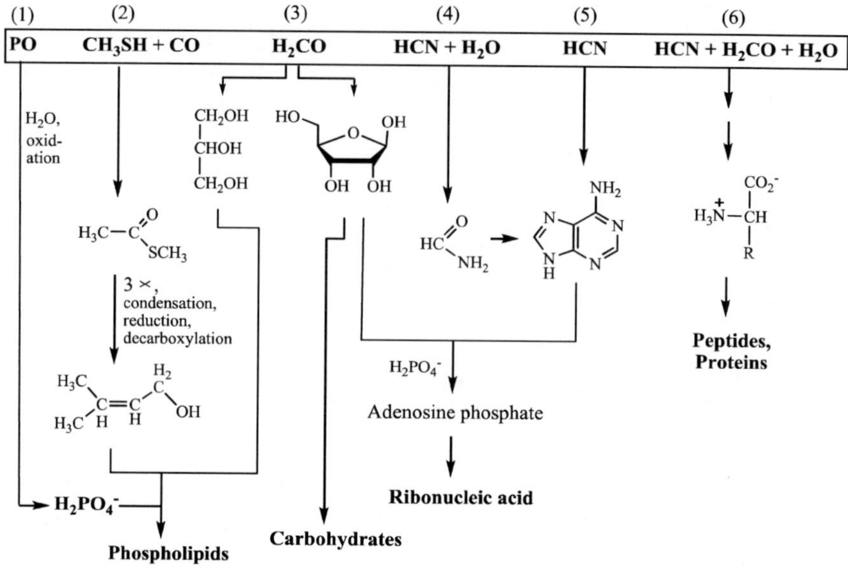

Figure 6.1. The putative formation of complex life molecules from basic interstellar precursors (upper row, framed): (1) Phosphorus oxide PO converts, in the presence of water and oxidation equivalents, to phosphate. (2) Methanethiol and carbon monoxide form the highly reactive methyl ester of thioacetic acid, three molecules of which make dimethylallyl alcohol. (3) Formaldehyde molecules can condense to form glycerol or ribose (formose reaction). Condensation of glycerol with dimethylallyl alcohol and phosphate yields phospholipids of the kind employed as building blocks in the membrane of archaea (**b** in Fig. 5.3). Polymerization by condensation of glucose molecules provides carbohydrates. (4) Hydrogen cyanide and water react to form formamide, one of the possible precursors for nucleobases such as adenine. (5) Adenine can also directly be generated from five molecules of hydrogen cyanide. Condensation of adenine, ribose and phosphate yields the nucleotide adenosine phosphate, and hence one of the building blocks of RNA (**a** in sidebar 5.1), the polymer for information storage in LUCA (Fig. 5.1) and many viruses, and for multiple tasks, including the transcription of information, in all more complex life forms. (6) Amino acids (pictured here in their zwitterionic form), and hence the modules that constitute peptides and proteins, are generated through the condensation of hydrogen cyanide, formaldehyde and water. Other precursor molecules for the formation of, e.g., glycine, have been proposed, such as $CO_2 + NH_3 + CH_2$; see also the main text for additional information on the genesis of amino acids.

6.1. THE GENERATION OF LIFE MOLECULES

Figure 6.1 provides an overview of those basic molecules present in interstellar clouds that can serve, once absorbed to the surface of dust grains, as starters for the creation of more complex molecules needed for the inventory of life. Of the ca. 180 molecules that have so far been detected in the interstellar medium, six are picked here: carbon monoxide CO, phosphorus monoxide PO, water H_2O, hydrogen cyanide (prussic acid) HCN, formaldehyde H_2CO, and methanethiol (mercaptane) CH_3SH. In the following, we will briefly address the formation of these molecules in the interstellar medium, and excursively outline how life molecules can basically develop from these simple precursors.

Several of the aspects of the creation of basic molecules in interstellar gas and dust clouds have already been discussed in some detail in section 3.2. Eqs. (6.1) to (6.5) are abbreviated representations of the interstellar formation of the six precursor molecules (framed in the top row in Figure 6.1) from basic ionic, atomic and molecular progenitors. The double arrows are used in Eqs. (6.1) to (6.5) indicate the *direction* of the reaction; the reactions are non-stoichiometric, nor are they balanced for charges; for more details (outlined in the previous section), see the links provided in parentheses.

$$\text{Phosphorus monoxide: } P^+ + OH \longrightarrow \longrightarrow PO \tag{6.1}$$

$$\text{Methanethiol: } C^+ + H_2 \longrightarrow \longrightarrow CH_2^+ \xrightarrow{S, H_2} \longrightarrow CH_3SH \quad \text{(details in sidebar 3.1)} \tag{6.2}$$

$$\text{Formaldehyde and carbon monoxide: } CH_2^+ \xrightarrow{e^-} \longrightarrow CH_2O; \ CO$$
$$H^+ + O^- \xrightarrow{\quad OH \quad} \quad \text{(see also sidebar 3.1)} \tag{6.3}$$

$$\text{Hydrogen cyanide: } N^+ + H_2 \longrightarrow \longrightarrow NH_3 \xrightarrow{C^+, e^-} \longrightarrow HCN \quad \text{(details in sidebar 3.1)} \tag{6.4}$$

$$\text{Water: } H + H + O \xrightarrow{\text{(dust)}} H_2O \longleftarrow \longleftarrow O^+ + H_2 + e^- \quad \text{(see Fig. 3.2 and Eq. (4.3))} \tag{6.5}$$

Typically, molecules in interstellar clouds form in the gas phase by ion-molecule reactions, i.e., in reactions where an effective cross section is secured by molecules possessing a dipole moment, such as OH, NH_3 and H_2O. Elemental hydrogen is present in the form of neutral atoms (H) or molecules (H_2), negative and positive ions (hydride H^- and protons H^+), and molecular cations (H_3^+). Dust particles can also be reaction partners, as shown for the formation of water in Eq. (6.5), and detailed in Figure 3.2 in section 3.1. Also abundant in the ice mantles of interstellar dust grains are methanol CH_3OH and OH, precursors for the formation of formaldehyde H_2CO, represented by Eq. (6.6), as an alternative to the generation of H_2CO from HCN + H_2O, entry (4) in Figure 6.1.

$$CH_3OH + {}^{\cdot}OH \rightarrow\rightarrow {}^{\cdot}CH_3O \rightarrow\rightarrow H_2CO \qquad (6.6)$$

The oxidation of CH_3OH by OH to form the methoxy radical CH_3O is noteworthy insofar as there is an activation barrier that should prevent progression of the reaction at the very low temperatures in interstellar clouds. This barrier can, however, be tunneled by the intermediate formation of a hydrogen-bonded complex $\{CH_3OH\text{---}OH\}$ which decomposes to form CH_3O (or CH_2OH) + H_2O (Shannon 2013). CH_3OH by itself can accrue in the outer ice layer of grains from CH_3 and OH radicals, Eq. (6.7), a reaction that is initiated by UV radiation.

$$ {}^{\cdot}CH_3 + {}^{\cdot}OH \rightarrow CH_3OH \qquad (6.7)$$

The reaction between methanol and the formyl radical HCO on grain surfaces yields glycolaldehyde $CH_2(OH)\text{--}CHO$, the simplest sugar, and an intermediate in the path towards more complex biologically relevant molecules. HCO in turn is generated from carbon monoxide and hydrogen atoms. The reactions are depicted in Eqs. (6.8) and (6.9) (Woods 2012). Glycolaldehyde has been detected in the outer spheres of a solar-type protostar (Jørgensen 2012).

$$CO + H \rightarrow HCO \qquad (6.8)$$

$$CH_3OH + HCO \rightarrow\ \rightarrow CH_2(OH)\text{--}CHO \qquad (6.9)$$

Another intriguing example in the context of precursors for life molecules, amino acids in this case, is the detection of aminoacetonitrile in the Large Molecule Heimat (Belloche 2008), a giant gas cloud near the galactic center. Aminoacetonitrile can be hydrolyzed to glycine, Eq. (6.10), and is thus a potential precursor for this simplest of the set of essential amino acids.

$$H_2N\text{--}CH_2\text{--}C\equiv N + 2H_2O \rightarrow H_2N\text{--}CH_2\text{--}CO_2H + NH_3 \qquad (6.10)$$

The laboratory scale synthesis of aminoacetonitrile, starting from ammonia, formaldehyde and hydrogen cyanide, Eq. (6.11a), the so-called Strecker synthesis, should only marginally contribute to the formation of amino acetonitrile in interstellar gas and dust clouds. An alternative route to amino acetonitrile is the reaction between methylene imine CH_2NH and hydrogen isocyanide HNC, Eq. (6.11b). This reaction is effectively catalyzed by water. Finally, UV-photolysis of a mixture of water, methanol, ammonia and hydrogencyanide yields, *inter alia*, acetonitrile, as shown in the non-stoichiometric reaction represented by Eq. (6.11c)

$$NH_3 + H_2CO + HCN \rightarrow H_2N–CH_2–C\equiv N + H_2O \qquad (6.11a)$$

$$CH_2NH + HNC \rightarrow H_2N–CH_2–C\equiv N \qquad (6.11b)$$

$$H_2O + CH_3OH + NH_3 + HCN \rightarrow \rightarrow H_2N–CH_2–C\equiv N \qquad (6.11c)$$

Yet another reaction path leading to the glycine precursor aminoacetonitrile is the radical-radical coupling between $^\cdot CH_2NH_2$ and $^\cdot CN$, Eq. (6.12).

$$CH_2NH_2 + CN \rightarrow H_2N–CH_2–C\equiv N \qquad (6.12)$$

Aminoacetonitrile can also be subject to hydrogen abstraction to form the radical $NH_2{}^\cdot CHCN$. This radical may then undergo a radical-radical reaction with $^\cdot CH_2X$, recovering aminoacetonitrile (X = H) or generating the nitrile precursors of serine (X = OH) and cysteine (X = SH), Eqs. (6.13a) and (13b) (Knowles 2010). From these nitriles, serine and cysteine are formed by reaction with water in analogy to the hydrolysis depicted in Eq. (6.10).

$$NH_2CHCN + CH_2OH \rightarrow NH_2CH(CH_2OH)CN \qquad (6.13a)$$

$$NH_2CHCN + CH_2SH \rightarrow NH_2CH(CH_2SH)CN \qquad (6.13b)$$

Laboratory experiments have also provided evidence for the *direct* formation of glycine either from methylamine and carbon dioxide by bombardment with high energy electrons, Eq. (6.14a), or from ammonia, CH_2 and carbon dioxide, Eq. (6.14b). In the latter case, an intermediate CH_2NH_3 (an isomer of methylamine CH_3NH_2) is generated.

$$(e^-) \ CH_3NH_2 + CO_2 \rightarrow H_2N-CH_2-CO_2H \qquad\qquad (6.14a)$$

$$NH_3 + CH_2 + CO_2 \ (\rightarrow H_3NCH_2 + CO_2) \rightarrow H_2N-CH_2-CO_2H \qquad (6.14b)$$

Glycine by itself has not been detected in the interstellar medium, likely reflecting glycine's fragility with respect to cosmic radiation. Glycine has, however, been found in meteorites and also in particles retrieved by the spacecraft Stardust from the coma of the comet Wild 2 (pronounced "vilt 2"). The Stardust mission started in 1999; the spaceship arrived at Wild 2 in 2004, where it collected samples from the comet's coma as well as dust particles from the interstellar space. Since returning to Earth in 2006, these samples have been analyzed, revealing comparatively high contents of silicates rich in magnesium and iron (Westphal 2014).

The formation of amino acids in interstellar clouds has been outlined here in some detail because amino acids are *the* building blocks for the bulk of organic matter constituting life. Additional scenarios for the generation of amino acids in non-interstellar environments, such as on other celestial bodies in our Solar System on the one hand, and the primordial Earth on the other hand, will be discussed in sections 6.2 and 6.3.

Yet another important issue in the context of life molecules is chirality. As mentioned earlier (see Figures 5.2, 5.3 and 5.4 in section 5.1.1), the chiral carbon (the carbon carrying the amino substituent) of amino acids in living organisms on Earth almost exclusively is in the L (or *S*) configuration, while the sugars ribose and deoxyribose are in the D configuration, and the central carbon of glycerol in lipids can be either L (eukarya and bacteria) or D (archaea).[24] The selection of amino acids in the L configuration during the primordial constitution of life was likely a random process, steered by a miticulous excess of the L over the D configurated form. But how did this tiny amplification of one of the isomers from a racemic[25] mixture come about?

It is well known that organic matter can be destructed by highly energetic UV light, so-called vacuum UV with wave lengths below 200 nm. Vacuum UV in space is commonly circularly polarized. The vector defining the strength and the direction of the electric field of an electromagnetic wave can be oriented either into one preferential direction, or the vector can describe a circle. The electromagnetic wave is then referred to as being either "linearly

---

[24] See footnote 15 in section 5.1.1 for the descriptors *R*, *S*; and D, L.
[25] A racemate is a 1:1 mixture of the two optical isomers of a chiral molecule.

polarized", or "circularly polarized", abbreviated to CPUV in the latter case. In circularly polarized light, the rotating vector describes a helix along the wave's direction of propagation (see Figure 6.2); the sense of rotation can be either left-handed (anti-clockwise) or right-handed (clockwise). As a racemic mixture of a molecule (here: an amino acid) interacts with CPUV, both enantiomers are gradually broken down. However, since one of the enantiomers absorbs CPUV with a slight preference over the other enantiomer, the breakdown rates of the two optical isomers differ. As a result, one of the two isomers becomes enriched in the mix of the molecules which survive the breakdown.

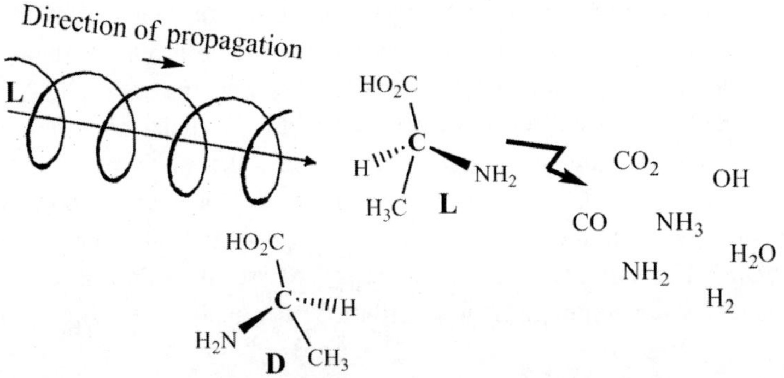

Figure 6.2. An illustration of the selective fragmentation of interstellar L-alanine by left-handed (L) circular polarized vacuum UV. The descriptors L and D are used here instead of the (for amino acids) more common descriptors S and R.

Chirality-dependent bond cleavage with concomitant destruction of a molecule can also come about by slow spin-aligned (spin-polarized) electrons, where "spin", descriptively speaking, refers to a rotation of the electron about its axis. This rotation can be left-handed (spin-up ↑) or right-handed (spin-down ↓), and goes along with a corresponding orientation of the magnetic moment. Electrons are abundant in the interstellar space; the spin-alignment is forced onto the electrons in the interstellar magnetic field. Again, one of the enantiomers of a chiral molecule is more receptive to the attack by the spin-polarized electrons and thus disassembled preferentially.

Further amplification of one of the optical isomers can occur as the molecules become absorbed to grains in comparatively dense interstellar clouds, and to aggregations of these grains to more massive particles that serve as building blocks of planetesimals, asteroids (the "leftovers" of planet formation) and comets. The most simplistic amplification comes about as the absorbing material is chiral by itself. Examples of such materials are certain clays (section 6.3) and quartz. In quartz, $(SiO_2)_\infty$, SiO_4 tetrahedra are linked together to helices which in turn are inter-linked to form three-dimensional structures. Such a helix can be left-handed or right-handed.[26]

Other mechanisms for the chiral amplification of molecules take place in contact with a solid/liquid interface. These mechanisms include asymmetric auto-catalysis, solid-liquid amplification, and aqueous alteration. In the case of asymmetric autocatalysis, a chiral molecule enantio-selectively catalyzes its own replication: the chiral molecule serves as a catalyst in the production of its siblings, at the same time suppressing the fabrication of its enantiomer. Solid-liquid amplification (Viedma 2008) presupposes the presence of a saturated solution of the enantiomers. If this solution is in equilibrium with its solid phase, one of the enantiomers preferentially goes into the solid phase, while the other one becomes enriched in the solute phase. Aqueous alteration comes in as a racemic mixture, integrated into a chiral mineral, is in contact with water: Selective binding of one of the enantiomers, for example an L-amino acid, to surface structures of the chiral mineral protects it from being washed out by water. The partially high enantiomeric excesses of amino acids found in some meteorites (section 6.2) are attributed to aqueous alteration (Glavin 2009).

6.2. DELIVERY OF ORGANIC COMPOUNDS BY METEORITES AND COMETS

In section 4.3, we have already briefly addressed meteorites as bringers of extraterrestrial matter to Earth. Meteorites are primordial agglomerates of dust and grains ("micrometeorites") from the very early phase of the formation of our Solar System, and in part even contain a record that can date back to the interstellar cloud prior to its contraction and the formation of the protoplanetary nebula. Meteorites can also be cut-offs of colliding asteroids, and consequently can provide insight into structural, mineralogical and chemical features intrinsic to the time when planetesimals, the progenitors of our planets, formed. Finally, a small but distinct number of meteorites originate from Mars and our Moon; and we have of course also available

---

[26] See also (**c**) in sidebar 5.1 for DNA as an example of a left-handed helix.

matter carried to Earth from the Moon by the Apollo mission in 1969 and, as mentioned above, from the coma of comet Wild 2.

A specific type of meteorites is represented by the carbonaceous chondrites (C chondrites) with up to about 3% carbon. The main share of this carbon, about 90%, is organic matter, one-third of which is of low molecular mass and soluble in organic solvents, while the remaining two-thirds are insoluble macromolecules similar to kerogens, i.e., a mix of straight-chain and branched-chain C_{10} to C_{16} paraffins, plus cyclo-paraffins (naphthenes). The remaining 10% of the carbon are apportioned to carbonates, silicon carbide SiC, and elemental carbon: graphite, diamond, lonsdalite (a carbon modification related to diamond), graphene (Two-dimensional layers of graphite), and others. To distinguish between terrestrial contamination and primordial origin of the carbon species, the isotopic ratios of stable isotopes are being consulted, in particular the ratios $^{2}H/^{1}H$ and $^{15}N/^{14}N$. An enrichment of ^{2}H with respect to ^{1}H, and ^{15}N with respect to ^{14}N, is a solid indication of a prebiotic origin: the low temperatures in the pristine gas and dust cloud that preceded our planetary system favor the retention of the heavier isotopes.

The delivery of organic compounds by carbonaceous chondrites to the early Earth is a potentially important source for Earth's prebiotic inventory. A well-documented example of a contemporary C chondrite is the Murchison meteorite (Glavin 2014) that fell in Western Australia in 1969. More than 80 amino acids have been identified in this meteorite, many of which are unknown in the terrestrial biosphere. For some of the indigenous amino acids in the Murchison meteorite (and a few other C chondrites as well), there is an enantiomeric excess *ee* of the L enantiomer, particularly distinct for isovaline, Fig. 6.3, with an *ee* of 18.5%. Isovaline of terrestrial origin is rare on our planet, and predominantly found in the D configuration. The enrichment in L-isovaline in the Murchison meteorite likely is a consequence of aqueous alteration (vide supra). The meteorite contains mineral hydrates; the water has supposedly been delivered to the meteorite's parent asteroid via the melting of ice inside the asteroid by "electrodynamic heating". Electrodynamic heating (Menzel 2013) comes about as the asteroid moves through the magnetic field of the Solar System, thus experiencing an electric field that heats the asteroid's interior. Alternatively, or additionally, radioactive heating comes in, for example in the course of the decay of ^{26}Al, Eq. (4.4) in section 4.2, or the decay of ^{60}Fe. ^{60}Fe is a β^{-} emitter that decays with a half-life of $2.6 \cdot 10^{6}$ years to ^{60}Co, Eq. (6.15).

$$\ce{^{60}_{26}Fe} \longrightarrow \ce{^{60}_{27}Co} + \ce{^{0}_{-1}e^-} + \bar{\nu}$$

$$(6.15)$$

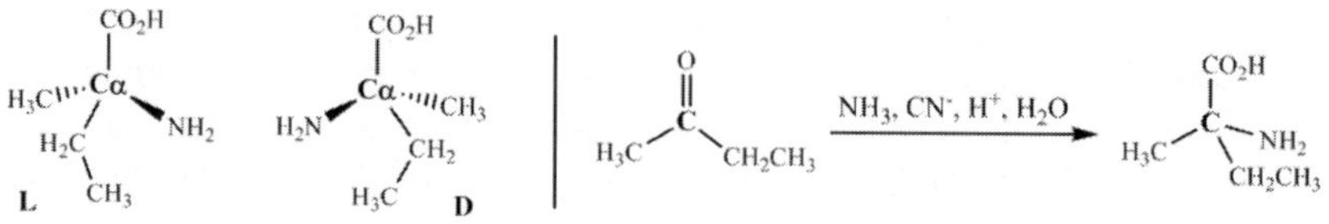

Figure 6.3. Left: The L (S) and D (R) enantiomers of isovaline. Cα is the center of the tetrahedron spanned by the four substituents. Right: The net reaction of the Strecker synthesis of isovaline, starting from 2-butanone.

Isovaline is not an essential amino acid for living organisms on our planet, in contrast to valine. In valine, the methyl substituent at the carbon Cα (i.e., the carbon adjacent to the carboxylic acid group) in isovaline is replaced by a proton; Cα thus is a *secondary* carbon, in contrast to the *tertiary* Cα in isovaline. Once a specific enantiomer is established for a chiral tertiary carbon center, it remains practically stable with respect to racemization for eons.

In contrast, chiral amino acids with a secondary Cα, such as the essential amino acids building up peptides and proteins, are too short-lived to allow for the survival of an excess of the L or D enantiomer over long periods of time; accordingly, more or less rapid racemization occurs. The survival of stable L-isovaline in the Murchison meteorite models a situation that might have prevailed some 3.8 billion years ago, when life began to develop on our planet, conceivably resorting to L amino acids provided from space by meteorites. A most probable precursor to isovaline is 2-butanone. In the presence of ammonia, cyanide and protons, 2-butanone reacts to form isovaline (so-called Strecker synthesis); this reaction is included in Figure 6.3.

Along with the common occasional meteorites that survive the passage through the atmosphere and can thus be collected and investigated, micrometeors (dust particles) constantly bombard our planet. The influx of dust particles originating from interplanetary and interstellar sources, with sizes up to 0.2 mm, amounts to at least 60 tons daily. Most of these particles are annihilated in the atmosphere by frictional heat, but a sizable amount experiences sufficiently effective deceleration to allow for the survival of organic matter (Coulson 2006).

A conceivable example of organics associated with dust is glycine, the simplest amino acid, the formation of which on the icy surfaces of dust grains has been discussed in section 6.1. Abiotic *macro*molecular carbonaceous matter similar to kerogen, likely stemming from micrometeorite impacts into lunar regolith and hence of pre-terrestrial origin, has also been detected in glassy silicate beads (Thomas-Keprta 2013) collected in the frame of the Apollo mission on the Moon.

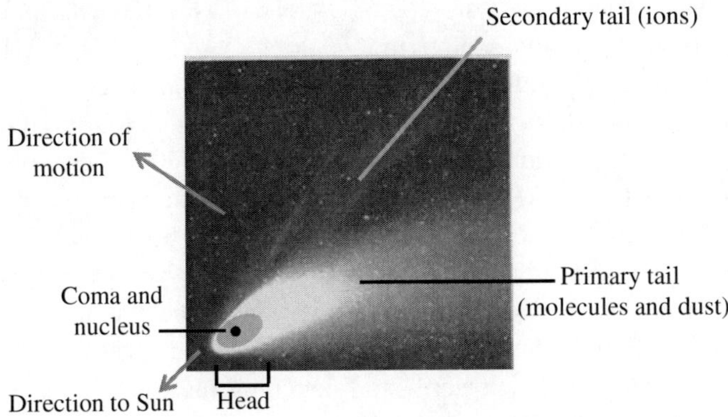

Figure 6.4. The comet Hale-Bopp (1995), and general cometary features. The coma and nucleus are not to scale.

Additional sources of molecules that can be basic building blocks for life are comets. Prior to their appearance as comets, i.e., as a celestial body with the typical tail(s) when approaching the Sun, these objects constitute a low density mixture ($\rho \approx 0.35$ g cm^{-3}) of ice and dust. Comets are therefore sometimes referred to as "dirty snowballs". The volatile icy components are dominated by water; the ^{1}H/^{2}H ratio for Kuiper belt comets corresponds to that of the water on Earth, suggesting that Earth's oceans have been replenished by impacts of comets to a significant extent. Cometary dust particles are in the micro- to centimeter range, and can be rich in carbonaceous components or silicates[27]. When drawing closer to the Sun, part of the comet's material evaporates and forms a coma, a kind of extended atmosphere, together with the tails (Figure 6.4): a shiny curved tail harboring molecular and dusty ingredients that are blown off the comet by the solar wind and left behind as the comet moves on its highly elliptic orbit, and a fainter, often bluish or mauve tail containing ions. This ion tail is oriented along the field lines of the Sun's magnetic field and thus points into the direction opposite the Sun. Life

---

[27] A comparatively large portion of the siliceous components is in the crystalline state, presupposing that these materials have formed close to the Sun, followed by rapid cooling down and transportation into the outer regions of the Solar System.

molecules such as glycine have been shown to be present in comets. More complex amino acids and, in addition, N-heterocycles can form from simple precursors molecules contained in comets by the energy engendered as comets crash into rocky formations of Earth (Martins 2013). Examples of precursor molecules detected in comets are CO, CO_2, H_2O, NH_3, CH_3OH and H_2CO. Impact events of this kind, delivering – along with water – life-building compounds to the prebiotic Earth, did occur billions of years ago.

Meteorites can also originate from our neighbor planet Mars. To date, the number of confirmed Martian meteorites is about 70. These meteorites represent fragments from Mars' rocky surface, ejected from the planet by spallation, i.e., when struck by a sufficiently energetic body, such as a small asteroid, at an angle allowing the knocked-out object to escape the gravitational force of the planet; Mars' gravitational acceleration is just about one-third that of Earth. The Martian origin of these meteorites is commonly ensured by analyzing gas inclusions, and matching these inclusions with the well documented composition of the Martian atmosphere. Carbon – in the form of carbonate, graphite and polyaromatic hydrocarbons – has, for example, been found in the Tissint meteorite that crashed down to Earth in the region of Morocco in 2011, or in the famous meteorite ALH84001, collected in Alan Hills, Antarctica, in 1984. While this simply tells us that there is carbon on Mars – certainly not a surprise – it has spurred speculations with respect to the presence of primitive life on Mars, or even an origin of life on Mars.

Three geological eras are commonly allotted to Mars in its development through the eons: The Noachian, the Hesperian and the Amazonian. The warm and wet Noachian period, lasting from 3.8 to 3.5 Gy, is characterized by high impact rates of meteorites, and presumably abundant surface water, the Hesperian (3.5 to 1.8 Gy) by widespread volcanic activity and perhaps catastrophic flooding. The Amazonian, starting about 1.8 Gy ago and lasting to the present, is the hyperarid and cold era that also shapes today's appearance of Mars. One of the tasks of the present Mars mission performed by the rover Curiosity, roaming around the approximately 3.8-year-old impact crater Gale near the equator since August 2012, is dedicated to the exploration of former or even actual primitive life on Mars.

The problem with detecting organics on Mars – or at least in the surficial areas of Mars' soil and rock that are accessible to probe sampling – is threefold: (i) Mars' surface is exposed to distinctive Sun-born and cosmic radiation as a consequence of the thin Martian atmosphere. The radiation absorbed dose on Mars' surface is 0.6 mSv per *day*, and hence approximately the *yearly* dose on Earth. (ii) The presence of sodium perchlorate $NaClO_4$ in Martian soils triggers the oxidative destruction of organics, but the oxidative chlorination of organics as well. Eqs. (6.16) and (6.17) are examples of methane as a putative substrate for oxidation and chlorination by chlorates. (iii) Carbon compounds are present in the rover as contaminants from Earth, complicating the identification of originary Martian carbon species.

$$NaClO_4 + CH_4 \rightarrow NaCl + CO_2 + 2H_2O \qquad (6.16)$$

$$NaClO_4 \rightarrow NaClO + 3/2O_2;\ 3NaClO + CH_4 + 3H^+ \rightarrow CHCl_3 + 3H_2O + 3Na^+ \qquad (6.17)$$

Recent evidence has, however, revealed that chlorinated organic compounds detected by Curiosity's spectrometric equipment chiefly came from organic matter in ancient Martian rock. These compounds include chlorinated methane, Eq. (6.16), ethane, propane, and likely also chlorobenzene. The possible origin of these compounds is either organics in cosmic debris roped in by Mars, or remains of Martian life from eons ago (Grotzinger 2014, Kerr 2014). The Gale crater on Mars in fact shows patterns reminiscent of ancient aqueous environments, including lakes with low salinity and ground water networks. Ancient clay-forming subsurface aqueous environments have previously also been detected by the Opportunity lander. The Yellowstone Bay, an approximately 5 m deep depression in Gale crater, contains layers of sediments hinting towards an ancient lake; these sediments still contain up to 3% water in the form of water of hydration. Another intriguing finding is subsurface *ferrous* (Fe^{2+}) iron – in contrast to the reddish surface *ferric* (Fe^{3+}) iron responsible for the red-tinged surface of Mars (and the planet's name: Mars is the Greek war god). The presence of ferrous iron indicates an environment devoid of oxidants such as surface perchlorates and atmospheric[28] O_2, and hence reducing conditions comparable to those of the primordial Earth.

All of these findings on Mars undeniably demonstrate that Mars could have been habitable in its early days. Microbes might have taken refuge in

---

[28] The mean surface level pressure of the Martian atmosphere is $6 \cdot 10^3$ Pa (6 mbar, which compares to the pressure in the terrestrial atmosphere at a height of 40 km). The main constituents are CO_2 (95.7%), N_2 (2.7%), Ar (1.6%) and O_2 (0.13%). Methane CH_4, the (seasonal) presence of which has repeatedly been reported and linked to, *inter alia*, methanogenic bacterial populations in subsurface lakes on Mars, has so far not been confirmed.

niches when Mars lost its dense atmosphere some 3 Gy ago, relying for example on the oxidation of H_2S as an energy source to sustain their life. The survival of primitive prokaryan life forms in subsurface niches wheresoever on Mars can thus not readily be ruled out. If so, this does not, of course, imply (nor exclude) that Mars actually had been inhabited by life forms resembling our terrestrial archaea and bacteria. In any case, the period of habitability did not last sufficiently long to allow for the evolution of more developed life forms, such as protozoa, algae and fungi. Primitive life forms (microbes) may well have been exchanged between Mars and Earth, transported from one planet to the other by rocks blasted off the surface in the course of impacts of asteroids.

The potential transport of primitive life forms between Mars and Earth, and the supply of organics to Earth by meteorites of asteroidal origin, by micrometeorites and comets, has spurred the notion of panspermia, a hypothesis according to which life is omnipresent in the Universe and has been (and is still being) distributed by meteoroids within solar systems, as well as by rogue planets and interstellar debris in between different solar systems. I will briefly come back to the ups and downs of this hypothesis, and the hypothesis' sensitivity and vulnerability in the context of (harmless) science fiction and (harmful) charlatanry in section 7.4.

6.3. PREBIOTIC SCENARIOS: CLAY ORGANISMS, MILLER-UREY, AND WÄCHTERSHÄUSER

Clays are layer minerals based on alumosilicates that, in the presence of water, form hydrogels with tiny empty spaces. This specific substructure favors the absorption and processing of chemical species and, at the same time, provides a protective confinement for biomolecules and biochemical reactions, very much in the way a cellular membrane of a living organism confines the inventory of cellular biomolecules (Yang 2013). Clays, and eventually other minerals with a particularly high surface-area-to-volume ratio, have possibly played an important role as templates for molecular species, protecting, selecting, concentrating, and catalyzing reactions of prebiotic organic molecules (Cleaves 2012), and are likewise of considerable interest in the context of the generation and protection of comparable organics on Mars and on rocky, water-bearing exoplanets: biomolecules such as amino acids, provided to Earth (and Mars) by meteorites, or formed in the primordial atmosphere or water reservoirs according to the Miller-Urey and/or Wächterhäuser and related scenarios (*vide infra*), readily attach to a clay surface and are thus protected from being damaged by diverse environmental factors. Such a microporous platform may also have served as a scaffold for the synthesis of more complex molecules; for example, the condensation of amino acids to peptides, or the formation of nucleotides from nucleobases + ribose/deoxyribose + phosphate, followed by further aggregation through condensation of these nucleotides to nucleic acids.

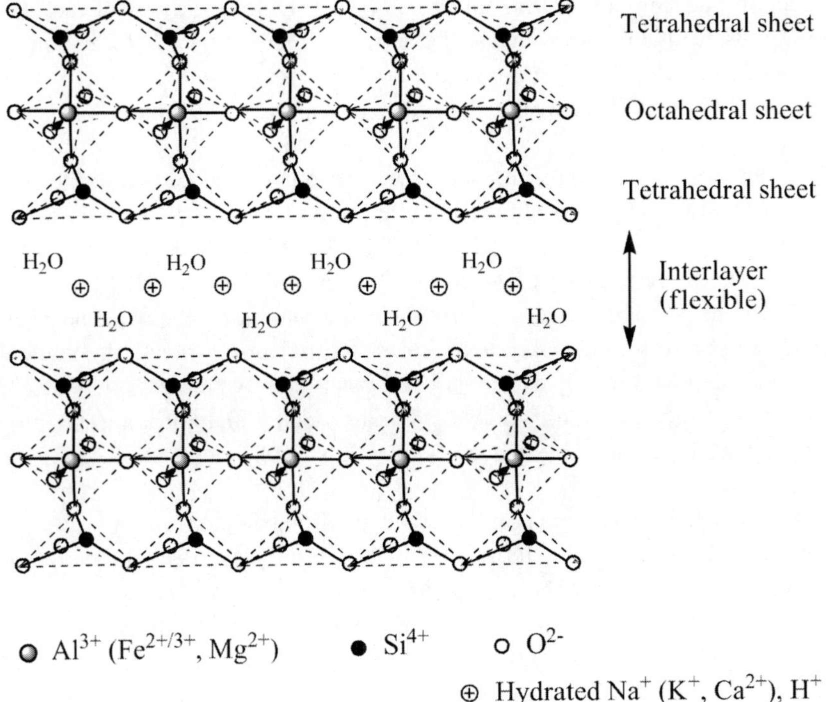

Figure 6.5. A schematic view of the clay mineral montmorillonite. In its "active" form (active with respect to the oligomerization of nucleotides), the interlayer contains [Na(H$_2$O)$_x$]$^+$ and/or protonated water (hydronium ions H$_3$O$^+$); the pH is slightly acidic to neutral. Montmorillonite type clays have been shown to catalyze the formation of peptides, nitrogen heterocycles (from formamide, which condenses into a variety of nitrogen heterocycles), nucleobases and lipid membranes.

This so-called "clay-hypothesis" (Cairns-Smith 2001) describes a possible process by which life arose from non-living matter – termed abiogenesis or biopoiesis – in a process catalyzed by clayey minerals. A typical clay material in this respect is montmorillonite (Figure 6.5). Montmorillonite is a phyllosilicate, where three-layered coatings of alumosilicates alternate with inter-layer spaces filled with mobile (partially) hydrated cations, Ca^{2+} and Na^+ in particular, and varying amounts of water. The layer spacing – 1 to 2 nm – is variable and depends on the amount of absorbed water.

Amino acids can be absorbed, and thus immobilized at the montmorillonite scaffold, both by ionic interaction between the carboxylate group of the amino acid and the Al^{3+} sites, and by hydrogen bonding interaction between amino acid functionality and O^{2-}/OH^- groups of the clay scaffold. These interactions reduce the activation barrier for the formation of the peptide bond, and thus promote the condensation of amino acids to form oligopeptides. Peptide formation can also be catalyzed by alumina (Figure (6.6a)).

In an acidic environment, interlayer cations in montmorillonite are replaced by H_3O^+. Montmorillonite modified in such a way has been shown to promote the oligomerization of RNA-nucleotides (Joshi 2009), and may thus have initiated information storage in the RNA world that preceded the DNA world. Up to 50-mers derived from adenosine and uridine have thus been generated (Figure (6.6b)). Clay hydrogel also efficiently protects the nucleic acids against breakdown by hydrolysis, and consequently preserves the nucleic acid's potentiality for transcription and translation. Montmorillonite is also present on Mars; comparable processes may thus also have taken place on our neighbor planet.

Although laboratory experiments with modified montmorillonite clearly demonstrated the axiomatic *possibility* of clay-catalyzed oligomerization of amino acids and of nucleobases, the *probability* for such a process to have occurred on the primordial Earth (or elsewhere in the Solar System and beyond) is exceedingly low, given the number of prerequisites that have to be fulfilled (Shapiro 2006). These prerequisites include (i) the availability of modified and thus active montmorillonite, (ii) a plausible reaction sequence that leads to the formation of the nucleotide in sufficiently large concentrations to allow for the formation of oligomers, and (iii) the absence of terminators of chain growth such as carboxylic acids, amines and molecules carrying alcoholic functions. Last but not least, it is very unlikely that a 50-mer randomly organized would act as a RNA replicase or a ribozyme for protein synthesis. Nonetheless, the clay organism approach remains attractive: Life *does* exist; hence, there had to have been a procedural flow by which modules allowing for the evolvement of life had been generated.

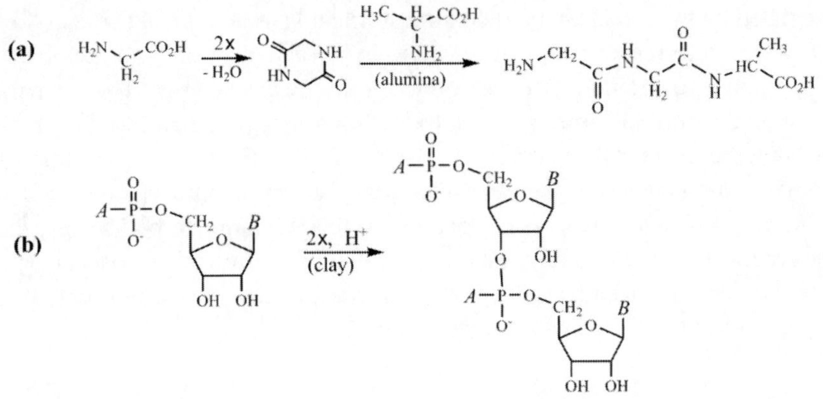

Figure 6.6. (**a**) Formation of the tripeptide alanyl-alanyl-valine on modified alumina (Al_2O_3). (**b**) Formation of a dinucleotide from activated nucleotides, catalyzed by a modified clay hydrogel such as montmorillonite (Figure 6.5). The letter *A* represents an activator (1-methyladenylium), and *B* is a nucleobase such as adenine and uracil.

In any case, prerequisites for the formation of more complex molecules such as peptides and nucleic acids include (i) the availability of simple precursor molecules, and (ii) the disposability of an energy source, since the formation of molecular organic units from basic inorganic precursors is an *endergonic*[29] process. Viable approaches in this context go back to Miller and Urey (Miller 1959) on the one hand, and to Wächterhäuser on the other (Wächtershäuser 2007). In the Miller-Urey and Wächtershäuser scenarios, the initialization of life starts at a lower level than in the clay organism approach, namely with the synthesis of low-molecular mass compounds via carbon fixation, resorting to the inorganic carbon sources CO and CO_2. Both gases have supposedly been present in the primordial atmosphere, along with N_2 and H_2O, as well as trace amounts of H_2 and O_2, and − at least locally − NH_3, HCN and H_2S.

---

[29] In an *endergonic* reaction (products → educts), the Gibbs free reaction energy ΔG is positive (> 0). Reactions with negative ΔG are termed *exergonic*. ΔG is composed of the change in reaction enthalpy (ΔH) and entropy ($T\Delta S$: T = temperature, S = entropy, ΔS = change in entropy) according to the Gibbs-Helmholtz equation $\Delta G = \Delta H - T\Delta S$. ΔH can be < 0 ($-\Delta H$, heat is released, *exothermic* process) or > 0 ($+\Delta H$, heat is consumed; *endothermic* process). Commonly, ΔH over-compensates $T\Delta S$, and an exothermic process (negative ΔH) is then also exergonic ($\Delta G < 0$). As, in an endothermic process ($\Delta H > 0$), the entropy drastically increases and $T\Delta S$ thus overcompensates ΔH, the chemical reaction becomes exergonic.

In the Wächtershäuser scenario, chemical evolution directed towards life is initiated by the autocatalytic redox transformation – on mineral surfaces – of CO_2 or CO into reactive organic species. "Autocatalytic" refers to the ability of a product to promote – once formed – its own synthesis. The primary catalysts are iron and iron-nickel sulfides, such as pyrite FeS_2, troilite FeS, pyrrhotite $Fe_{1-x}S$ (x < 0.2), greigite Fe_3S_4, pentlandite $(Fe,Ni)_9S_8$, and violarite $FeNi_2S_4$. Simultaneously, these sulfides can also act as redox partners. Eqs. (6.18) and (6.19) are two basic reactions in this scenario. In both cases, the reactive thioacetic acid methyl ester $CH_3C(O)SCH_3$ is formed. Thioacetic acid methyl ester is a constituent of acetyl coenzyme A, and hence a molecule that plays a pivotal role in many biochemical reactions.

$$2\ CH_3SH + CO \xrightarrow{(NiFeS)} CH_3-C\underset{SCH_3}{\overset{O}{\lessgtr}} + H_2S \tag{6.18}$$

$$2\ CH_3SH + CO_2 + FeS \longrightarrow CH_3-C\underset{SCH_3}{\overset{O}{\lessgtr}} + H_2O + FeS_2 \tag{6.19}$$

The starting product methanethiol (methylmercaptane) CH_3SH in Eqs. (6.18) and (6.19) is generated from hydrogen sulfide H_2S and CO or CO_2, Eqs. (6.20) and (6.21). The sulfide in FeS can act as the source for the electrons: sulfide S^{2-} becomes oxidized to disulfide S_2^{2-}, as shown in Eq. (6.21). Eqs. (6.20) and (6.21) represent the *net* reaction.

$$CO + H_2S + 4H^+ + 4e^- \rightarrow CH_3SH + H_2O \tag{6.20}$$

$$CO_2 + 3FeS + 4H_2S \rightarrow CH_3SH + 3FeS_2 + 2H_2O \tag{6.21}$$

In addition to being the catalyst and/or reaction partner, the sulfidic mineral furnishes shelter for the formed products insofar as it absorbs, and thus secures and accumulates, the newly-formed organic molecules in tiny niches in the surface realm. The absorption proceeds via coordinative bonds to the iron centers, hydrogen bonds, and electrostatic interaction; the inventory is thus protected against drifting and degradation. Micro-textured iron sulfides (Figure 6.7a) deposited in the proximity of hydrothermal vents can be particularly active in this respect.

The micrometer-sized pockets are sometimes referred to as "iron sulfide cells", and the overall assemblies have been dubbed "pioneer organisms"

(Wächtershäuser 2007) since they fulfill several of the criteria associated with the concept of life (cf. also section 5.2): (i) metabolism (chemical reactions); (ii) determinism (only a restricted number of synthetic paths is available); (iii) self-replication (the auto-catalytic feed-back); and (iv) evolution – in terms of variances and modifications in the course of chemical reactions.

It should be noted, however, that mineral surfaces with redox-active metal ions can also initiate the *degradation* of organic molecules in consequence of the formation of reactive oxygen species such as the superoxide radical, Eq. (6.22)

$$Fe^{2+} + O_2 \rightarrow Fe^{3+} + {}^{\cdot}O_2^{-} \tag{6.22}$$

The water phase surrounding these pioneer organisms is the source for "nutrients", i.e., for the basic molecules afforded for the synthesis of more complex ones. Interestingly, *all* living organisms on Earth invariably have available iron-sulfur cofactors in a large variety of redox-active enzymes. An example is shown in Figure 6b: a so-called ferredoxin based on four iron centers (linked to the protein matrix via cysteinate residues) and four sulfide centers. In contrast, however, to the iron sulfides involved in reactions such as represented by Eq. (6.21), where sulfide is the redox-active part ($2S^{2-} \rightarrow S_2^{2-} + 2e^{-}$), ferredoxins deliver or consume electrons by switching between Fe^{2+} and Fe^{3+}.

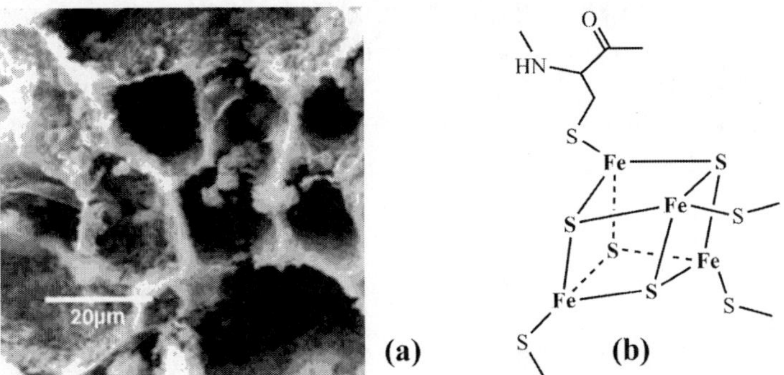

Figure 6.7. (a) Honey-comb structure of (artificially prepared) iron sulfide FeS as a model for FeS generated in the environment of deep sea hydrothermal vents (for hydrothermal vents see Fig. 5.11 in section 5.4). Reproduced by permission (Martin 2003); © Royal Society of Chemistry. (b) The catalytically active core of a [4Fe,4S]-ferredoxin. Each $Fe^{2+/3+}$ is linked to three trebly bridging S^{2-} and an end-on cysteinate(1–).

At the beginning of the second half of the past century, there was general consensus that the primordial atmosphere was a reducing one, consisting of H_2, N_2, CH_4, H_2O and NH_3 plus minor amounts of CO, CO_2, SO_2 and NO/NO_2. From a contemporary point of view, trace amounts of O_2 and H_2S (and PH_3) are believed to have additionally been present. Miller and Urey based their initial experiments on a gas mix of H_2, H_2O, CH_4 and NH_3 (Urey 1952, Miller 1953, Miller 1959), supplemented in consecutive experiments, also carried out by various other groups, with CO and CO_2, N_2, O_2, H_2S and other gases. Energy sources for activating and disrupting these atmospheric constituents for successive recombination to form more complex organic molecules are vacuum UV, electrical discharges (lightning, spark discharge), and to some extent also cosmic rays (essentially protons), and γ rays emitted in the frame of radioactive decay.

In their originary experiments, Miller and Urey clearly showed that a gas mix mimicking the primordial atmosphere, after being subjected to a spark discharge, contains various amino acids, including glycine, α-alanine and β-alanine, in addition to lactic acid, formic acid, acetic acid and urea. An intermediate product in the formation of glycine supposedly is aminoacetonitrile (which has also been detected in space; section 6.1), a view which is supported by complementary experiments: As (acidified) clay is present in the reaction vessels, the intermittently generated acetonitrile oligomerizes to form glycine peptides (Akabori 1956), linking to the clay organisms addressed above. For the generation of aminoacetonitrile according to the Strecker synthesis cf. Eq. (6.11a) in section 6.1, for its hydrolysis Eq. (6.10). The starting products in the Strecker synthesis, formaldehyde CH_2O and hydrogen cyanide HCN, arise from gaseous constituents that had been available in the primordial atmosphere, namely methane plus oxygen atoms (formed through water splitting), Eq. (6.23), and methane plus ammonia, Eq. (6.24).

$$CH_4 + 2O \rightarrow CH_2O + H_2O \tag{6.23}$$

$$CH_4 + NH_3 \rightarrow HCN + 2H_2 \tag{6.24}$$

Whatever initiated the creation of complex molecules and their development and organization into primitive life forms, starting approximately 4 to 3.8 billion years ago, there remains the intriguing question as to how these life forms could have coped with the "faint young Sun" problem. With the beginning of the archaean period (lasting from 3.8 to 2.5 Gy), the Sun's brightness would have been just about one-third of what it is today, because the nuclear reaction providing the main part of the Sun's energy, the core fusion of hydrogen to form helium, had not yet reached its magnitude at a later time. Earth should thus have been frozen over in most of its parts, enabling the development of cryophiles only, or restricting development of life to environmental niches warmed by volcanism. But the Sun's comparatively low luminosity may also have been counteracted by particularly high atmospheric concentrations of greenhouse gases, and/or a lower albedo than today, owing to the sufficiently more extended oceans and diminished cloud formation at that early time (Rosing 2010).

SUMMARY

This chapter summarizes possible scenarios for the generation of basic (organic) molecules at conditions prevalent in dense interstellar gas and dust clouds on the one hand, and in primordial planetary atmospheres, aquifers and porous minerals on the other hand. In addition, ways are indicated along which more complex molecules can form, molecules that are modules for an RNA world or a protein-first world.

Of the ca. 180 atoms, ions and molecules in interstellar clouds, six molecules have been selected that can be considered as progenitors for constituting more complex building blocks: PO, CO, H_2O, HCN, H_2CO and CH_3SH. These basic molecules can further develop into phosphate, lipids, carbohydrates, peptides and nucleic acids – via intermediates such as glycolaldehyde and aminoacetonitrile, detected in solar-like protostars and molecular clouds, respectively. All these six basic precursors for biomolecules are generated in interstellar clouds from basic atoms, ions and molecules such as P^+, OH, CH_2^+, H^+, O, N^+, H_2, C^+ and electrons. Particularly efficient in this respect are reactions in the surface ice mantles of dust grains.

Life as we know it falls back on chiral amino acids (the L enantiomers) and sugars (the D enantiomers). Chiral amplification can occur in the interstellar space by circular polarized UV, and as racemic mixtures come into contact with aqueous-solid interfaces in (chiral) minerals. An example of the latter is the enrichment of L-isovaline in the Murchison meteorite. Meteorites (in most cases fragments stemming from asteroids), along with micrometeorites and comets, have contributed to our planet's supply of life molecules. Recent clues to the presence of chlorinated hydrocarbons on Mars hint towards an input of carbon-based progenitor molecules, and thus molecules that are the basis of life. Together with evidence for water flows on the early Mars, this indicates that our neighboring planet may well have been habitable for microbes at least at about the time when life also evolved on Earth. The basic molecules for the synthesis of more complex ones, amino acids in particular, are not only potentially provided from space, but can also be synthesized from the constituents in a rocky planet's primordial atmosphere (Miller-Urey scenario).

Clay minerals are particularly apt for the formation, as well as the protection, of complex life molecules such as the oligomers of nucleotides (primitive RNAs) and amino acids (peptides); these minerals have therefore been referred to as clay organisms. On the other hand, freshly precipitated iron and iron-nickel sulfides both enable the formation of the highly reactive methyl ester of thioacetic acid (an analog of acetyl coenzyme-A) from methyl sulfide and CO or CO_2 (Wächtershäuser scenario). More generally, absorption phenomena on mineral surfaces, such as provided by clays and iron sulfides, could have assisted the replicator-first scenario (RNA world) as well as the metabolism-first scenario (protein world). To some extent, these scenarios are reminiscent of the Gaia[30] hypothesis, according to which organisms interact with mineral surroundings, thus creating a self-regulatory system that ascertains maintenance of the conditions for life.

REFERENCES

Akabori, S., K. Okawa, M. Sato (1956): Introduction of side chains into polyglycine dispersed on solid surface. *Bull. Chem. Soc. Japan* 29:608-611.

Belloche, A., K.M. Menten, C. Comito, et al. (2008): Detection of amino acetonitrile in Sgr B2(N). *Astronom. Astrophys.* 482:179-196.

Cairns-Smith (2001): The origin of Life: Clays. *Frontiers of Life* 1:169-192.

Cleaves, H.J. II, A.M. Scott, F.C. Hill, et al. (2012): Mineral-organic interfacial processes: potential roles in the origins of life. *Chem. Soc. Rev.* 41:5502-5525.

Coulson, S.G. (2006): Meteoroids: a landing capsule for panspermia. *Int. J. Astrobiol.* 5:307-312.

---

[30] The deity Gaia is the personification of Earth in Greek history. The Gaia hypothesis goes back to James Lovelock, and is discussed in more detail in the epilogue (ch. 8).

Glavin, D.P., J.P. Dworkin (2009): Enrichment of the amino acid L-isovaline by aqueous alteration on CI and CM meteorite parent bodies. *Proc. Natl. Acad. Sci.* 106:5487-5492.

Joshi, P.C., M.F. Aldersley, J.W. Delano, and J.P. Ferris (2009): Mechanism of montmorillonite catalysis in the formation of RNA oligomers. *J. Am. Chem. Soc.* 131:13369-13374.

Jørgensen, J.K., C. Favre, S.E. Bisschop, et al. (2012): Detection of the simplest sugar, glycolaldehyde, in a solar-type protostar with Alma. *Astrophys. J. Lett.* 754:L4.

Knowles, D.J., T. Wand, J.H. Bowie (2010): Radical formation of amino acid precursors in interstellar regions? Ser, Cys and Asp. *Org. Biomol. Chem.* 8:4934-4939.

Martin, W., M. Russels (2003): On the origins of cells: a hypothesis for the evolutionary transitions from abiotic geochemistry to chemoautotrophic prokaryotes, and from prokaryotes to nucleated cells. *Phil. Trans. Roy. Soc.* 358:59-85.

Martins, Z., M.C. Price, N. Goldman, et al. (2013): Shock synthesis of amino acids from impacting cometary and icy planet surface analogues. *Nature Geosci.* 6:1045-1049.

Menzel, R.L., W.G. Roberge (2013): Reexamination of induction heating of primitive bodies in protoplanetary disks. *Astrophys. J.* 776:89.

Miller, S.L. (1953): A production of amino acids under possible primitive earth conditions. *Science* 117:528-529.

Miller, S.L., H.C. Urey (1959): Organic compound synthesis on the primitive Earth. *Science* 130:245-251.

Rosing, M.T., D.K. Bird, N.H. Sleep, and C.J. Bjerrum (2010): No climate paradox under the faint early Sun. *Nature* 464:744-747.

Shannon, R.J, M.A. Blitz, A. Goddard, et al. (2013): Accelerated chemistry in the reaction between the hydroxyl radical and methanol at interstellar temperatures facilitated by tunneling. *Nature Chem.* 5:745-749.

Shapir, R. (2006): Small molecule interactions were central to the origin of life. *Quart. Rev. Biol.* 81:105-125.

Thomas-Keprta, K.L., S.J. Clemett, S. Messenger, et al. (2013): Organic matter on the Earth's Moon. *Geochim. Cosmochim. Acta* 134:1-15.

Urey, H.C. (1952): On the early chemical history of the Earth and the origin of life. *Geophysics* 38:351-361.

Viedma, C., J.E. Ortiz, T. de Torres, et al. (2008) Evolution of solid phase homochirality for a proteinogenic amino acid. *J. Am. Chem. Soc.* 130:15274-15276.

Wächtershäuser, G. (2007): On the chemistry and evolution of the pioneer organism. *Chem. Biodivers.* 4:584-602.

Westphal, A.J., R.M. Stroud, H.A. Bechtel, et al. (2014): Evidence for interstellar orgin of seven dust particles collected by the Stardust spacecraft. *Science* 345:786-791.

Woods, P.M, G. Kelly, S. Viti, et al. (2012): On the formation of glycolaldehyde in dense molecular clouds. *Astrophys. J.* 750:19.

Yang, D., S. Peng, M.R. Hartman, et al. (2013): Enhanced transcription and translation in clay hydrogel and implications for early life evolution. *Scientific Reports* 3:3165.

Chapter 7

HABITABILITY AND BIOSIGNATURES

There are varied paths for the synthesis of basic molecules necessary for life, and for the organization and arrangement of these basic building blocks into more complex entities and molecular assemblies that constitute the material nature of life. These pathways to more complex entities can be undirected, resulting in a multitude of disposable molecules, or partially directed, supported for example by substructures on mineral surfaces (section 6.3). Such "template-substrate pairings" (Gollihar 2014) should give rise to targeted selectivity within the product spectrum, thus allowing for the formation of more organized assemblies of substrates and hence a more efficient provision of organizational units to be amplified, and fed into feedback cycles in "metabolism-first" as well as in "replication-first" models for propagating chemical information content and thus sparking life.

Roughly 20% of Sun-like stars (early G stars) are believed to have approximately Earth-sized planets that potentially could provide the basis for the development of life. Perhaps even more suitable for initiating, evolving and harboring life are planetary systems around K stars (cf. sidebar 2.2 for the classification of stars), and hence stars that are slightly cooler and smaller than the Sun. So far, almost all of the potentially habitable planets have in fact been found in orbits around K stars.

In this chapter, I will first delineate the concepts "habitability" and "uninhabitability", and then resort to water as *the* element of life, in particular in celestial bodies beyond Earth in our Solar System, and here with special reference to the moons with sub-surface oceans. In this context, alternatives to water as the elixir of life will also be recaptured, i.e., the chemistry in Titan's atmosphere and liquid hydrocarbon oceans is illuminated. Briefly, the concept "panspermia" will then be addressed.

7.1. Habitability

The overall course of the development of basic molecules into Life is commonly considered to presuppose moderate temperatures (moderate in a sense that allows organic molecules to persist), and to require timescales of several hundred million years at least. The first prerequisite – moderate temperatures – was met merely 10-17 million years after the Big Bang (Loeb 2014), the event at which a space-time singularity (the initial state of the Universe; see sidebar 7.1) became resolved and developed into matter, time and space. The moderate temperatures between 273 and 373 K for this timespan, covering approximately seven million years, were maintained by the cosmic microwave background, and allowed liquid water to exist – and thus to establish and to sustain life as we know it[31] – on the surface of rocky planets orbiting the first massive stars that could have formed out of the extremely dense primordial matter.

There are, however, a couple of caveats rendering such a view a rather hypothetical one: The period during which this uniform warmth existed lasted a few million years only, hardly allowing for the development of even primitive life forms such as those represented by archaea and bacteria. Also, since the processes leading to the formation of life molecules are commonly exothermic reactions, a required energy sink would not have been available in the uniformly warm early Universe. In any case, life – if formed during the early period of the Universe – would hardly have survived the particularly high temperatures and highly energetic particle streams that accompanied the subsequent expansion and development of the Universe, nor the rapid death of these early planetary systems with their short-lived giant suns.

***

Sidebar 7.1. The development of the Universe

Our Universe came into existence 13.79 billion years ago, starting with the "Big Bang". The following periods of development can be recorded:

---

[31] For potential life forms resorting to liquids other than water see section 5.3 and also the implementation re prebiotic chemistry on the moons in our Solar System, Titan in particular, dealt with in section 7.3.

a) The *Planck Epoch*, lasting from 0 (zero) to 10^{-43} s, is characterized by a temperature $T = 10^{32}$ K and a density $\rho = 10^{94}$ g cm^{-3}. All of the four fundamental forces (the electromagnetic force, the strong nuclear force, the weak nuclear force, and the gravitational force) remained unified.

b) The *Grand Unification Epoch*, from 10^{-43} to 10^{-36} s. The gravitational force became decoupled from the three other forces. Electromagnetism, strong interaction and weak interaction were unified as the electronuclear force.

c) *Cosmic Inflation*, lasting from 10^{-36} to 10^{-32} s. Within this time span, the Universe expanded by a factor of 10^{50}. The temperature dropped to 10^{27} K.

d) The inflation period was followed by a slowed-down expansion of the Universe, starting with the *Hadron Epoch*: The strong force (which sticks together quarks) separated from the weak force (responsible for the radioactive decay) and the electromagnetic force. Quarks and antiquarks formed, and thus the building blocks of protons and neutrons (baryons, consisting of 3 quarks) and of hadrons (consisting of 2 quarks), along with the gluons, the particles that "glue" together protons and neutrons in the nucleus. This process proceeded within a time span of 10^{-12} to 10^{-6} s. The temperature dropped to 10^{13} K. Along with the baryons, *anti*baryons formed, though to a somewhat lesser extent. At the end of the Hadron Epoch, baryons and antibaryons started to anneal each other, leaving behind high-energy photons and a minute excess of baryons.

e) In the *Lepton Epoch* (10^{-2} s, 10^{12} K, $\rho \approx 10^{13}$ g cm^{-3}), the photons created in the Hadron Epoch collided to generate electrons and positrons (anti-electrons), with the density of electrons slightly exceeding that of the positrons.

f) In the time span 10^{-2} s to 3 min, *nuclear fusion* and thus the generation of the nuclei of deuterium (^{2}H), tritium (^{3}H), helium (^{3}He, ^{4}He) and also some lithium (^{7}Li) and beryllium (^{7}Be) took place.

g) 379,000 years later, the Universe had cooled down sufficiently, to about $3 \cdot 10^3$ K, to allow for the formation of neutral atoms that no longer absorb photons. Concomitantly, the Universe became transparent. This process also marks the manifestation of the cosmic background radiation that, due to the red-shift of this primordial light, is now observed as a background microwave radiation ($\lambda = 1.9$ mm) corresponding to a temperature of 2.725 K.

The oldest cosmic object so far detected is the galaxy z8_GND_5296, in the constellation Ursa Major, at a redshift $z = 7.51$ (see below for the definition of z), corresponding to an age of birth of 700 million years after the Big Bang, hence at a distance of 13.1 billion light years (Finkelstein 2013). The star formation rate in this old galaxy is 350 $m_\odot$ per year, i.e., about 100 times the rate in the Milky Way.

The redshift z is defined by $z = (\lambda_{observed} - \lambda_{emitted})/\lambda_{emitted}$, where λ is the wave length of the light. The redshift comes about by the Doppler effect: since the Universe is persistently expanding, the light emitted by the stars, as detected in our system, is shifted towards longer wavelengths, i.e., towards "red". The redshift z of the background microwave radiation is 1089, corresponding to $3.8 \cdot 10^3$ years after the Big Bang. For the time span $10 \cdot 10^6$ to $17 \cdot 10^6$ years (with temperatures that inherently would have provided habitability), z is in the range 137–100.

***

The habitability of a planet (or a moon) does not necessarily imply that the planet (or the moon) is in the stellar habitable zone; nor is habitability excluded for planets, and moons in particular, that are outside the stellar habitable zone. We have already pointed to this possibility in the context of rogue (or Steppenwolf) planets in sections 2.2 and 3.1. Moons may also come in: Titan, Saturn's largest moon (and the second-to-largest and most massive moon in our system) is an example of a moon with a dense atmosphere, about 1.5 as dense as Earth's (section 7.3). Likewise, the subsurface oceans on moons such as, in the Solar System, Titan, Europa and Enceladus are potential habitats; and – as noted in section 2.2 – the tidal heating of a moon by its planet can also provide habitability.

Such objects may even be more hospitable than Earth is; they are therefore sometimes called "superhabitable" (Heller 2013). Planets that are slightly more massive and older than Earth, and circling K stars (suns somewhat less massive than our Sun) have been proposed as candidates for superhabitability. On the other hand, planets in the stellar habitable zone can turn out to be uninhabitable (*vide infra*). In either case, habitability outside the habitable zone and uninhabitability inside the habitable zone can be caused by tidal heating. Tidal heating refers to the frictional dissipation of energy coupled to the orbital and rotational motions of the sun + planet, or planet + moon, and thus adds to the energy supplied through stellar radiation: the eventually hot planetary core, and the energy provided by radioactive decay.

Uninhabitability can be caused by a loss of obliquity and/or by the tidal locking of a planet (Heller 2011). A loss of obliquity, i.e., of the axial tilt, results in hot equator zones and cold polar regions, and thus in the loss of seasons to equilibrate temperatures. In the case of tidal locking, where the rotation period equals the orbital period (Mercury in our system is an example), the planet's side facing the sun is permanently irradiated and therefore particularly hot, while the opposite side is dark and cold. In that case, "re-equilibration" of the atmosphere can come in: a loss of the atmosphere by evaporation into the space, and by freezing out in the planetary regions that remain hidden from the sun. Most of all, dwarf stars quickly erase the axial tilt of Earth-sized planets and super-Earths.

Terrestrial planets orbiting Sun-like stars are sufficiently less susceptible to an erasement of the axial tilt.

How, apart from orbit and rotational features, can we obtain information on the potential habitability of an exoplanet? A pioneering future method in this respect is the spectropolarimetric analysis of star light reflected from an exoplanet. Such an analysis permits us to gather information on exoplanets; in particular, the exoplanet's atmospheric and surface features such as clouds and oceans, but also potential biosignatures arising from vegetation. Spectropolarimetry is a technique that provides, in addition to the intensity and spectral composition of a light source, also the light's state of polarization. The respective analysis of Earthshine (sunlight reflected by Earth to the Moon, and re-reflected from the Moon to Earth) reveals features from Earth's atmosphere and surface in the *linearly* polarized regime and from areas of vegetation by the degree of *circular* polarization[32] (Keller 2012).

7.2. WATER

At low pressure, typically in the interstellar space, water can only exist in its gaseous form, or confined to dust particles, in the form of ice; the sublimation temperature in common interstellar clouds with densities between

---

[32] The terms "linear polarization" and "circular polarization" have been explained in section 6.1.

10^3 and 10^6 atoms/molecules cm^{-3} is around 100 K. At a pressure > 600 Pa (the surface pressure on Mars), and a temperature above 273 K, water is liquid and in equilibrium with its gaseous state of aggregation (vapor). Metastable bulk liquid water can exist down to 227 K, and any atomic, molecular and ionic components dissolved in water depress the freezing point, in extreme cases, such as sodium perchlorate ($Na^+ + ClO_4^-$) dissolved in water, down to ca 200 K; see also section 5.3. Also, high pressure reduces the freezing point; note that ice is *less* dense than liquid water with a maximum density of, by definition, 1 g ml^{-1} at 4 °C. Finally, solid water can form clathrate hydrates, where polar molecules such as NH_3 and H_2S as well as non-polar molecules such as CO_2 or CH_4 are trapped within the crystalline framework of water molecules (see Figure 5.9 in section 5.3).

In addition to "normal" water H_2O, the "heavy" (deuterated) forms of water HDO and D_2O exist along with water where the common oxygen isotope ^{16}O is replaced for ^{17}O or ^{18}O. Deuterium enrichment in water is particularly effective in gas-grain processes in interstellar clouds. Finally, water is commonly present as a mix of *para-* and *ortho-*H_2O, distinct by the antiparallel (*para*) and parallel orientation (*ortho*) of the spins of the two hydrogen atoms. This is illustrated in Fig. 7.1, showing a section of the far infrared spectrum of the disk of dust and water vapor surrounding the dying star W Hydrae (Barlow 1996).

The fundamental importance of water for all life as we know it has repeatedly been emphasized in the preceding chapters, including a detailed description of those properties of water that are related to its particularly well-adjusted characteristics as it comes to the generation of life molecules (section 6.1) and the sustenance of life (section 5.3). The omnipresence of water throughout our planetary system and in other stellar systems has been noted, and shall briefly be summarized and supplemented.

In the present day Solar System, marked reservoirs of water ice are particularly abundant in the outer asteroid belt and in the Kuiper and scattered disk belts. Trapped water, i.e., H_2O confined to minerals such as apatite, olivine and plagioclase in the form of hydrates or hydroxides, is present on the moon (Anand 2014). Ringwoodite, a high pressure modification of olivine in the depths of Earth's crust, the transition zone, is possibly an equally important source of water on our planet, in addition to the reservoirs of unbound water (section 4.2).

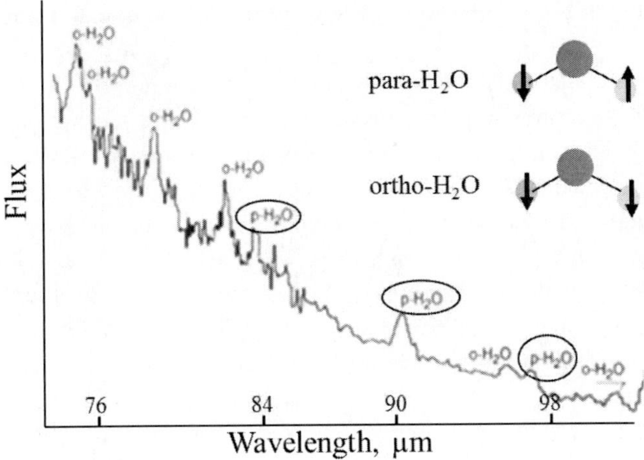

Figure 7.1. Section (far infrared) of the spectrum of the circumstellar disk (dust and water) of the AGB star W Hydrae, showing the water vapor lines of *para-* and *ortho-* water, p-H_2O (circled) and o-H_2O. Reproduced with permission from Mike Barlow, University College London. The arrows in the water molecules indicate the spin direction (antiparallel vs. parallel) of the two protons.

There are subsurface oceans on the moons Titan, Enceladus and Europa. Subglacial aquatic environments beneath the West Antarctic ice sheet, devoid of light and with temperatures slightly below zero, thrive with bacteria and archaea (Christner 2014). These microbial systems may well be considered to model respective situations in the subsurface oceans of these moons. The largest moon in our Solar System, the Jupiter moon Ganymede, is composed of approximately equal amounts of water ice and siliceous minerals/rock, a composition that also pertains to Jupiter's Callisto. Saturn's rings consist of chunks of ice, and water ice is also abundant on the dwarf planets beyond Uranus. All of these objects are on the far side of the so-called snow-line (or frost-line)[33], where water-ice deposited on and in grains of the protoplanetary nebula would have survived. But, as we are aware, water is also present on objects *inside* the snowline; our home planet is a prominent example, and the Mars rovers Opportunity and Curiosity have traced signs of ancient water on Mars (cf. also sections 5.3 and 6.2). Water may in fact still be around on the

---

[33] The present snow-line is at about 2.7 AU, i.e. close to the outside margin of the asteroid belt. In the protoplanetary disk, the snow-line was positioned at a distance of ca. 5 AU.

present Mars, at least in the form of ice and liquid perchlorate brines (Fischer 2014).

Water ice has also been detected near Mercury's north pole in permanently shadowed regions, mostly layered by organic lag deposits, and thus strongly indicating that comets are a potential source of these ice deposits (Lucey 2013, Chabot 2014). There are, however, also rare spots of unlayered ice that might originate from solar wind protons interacting with oxidic minerals on Mercury's surface (see Eqs. (4.2) and (4.3) in section 4.2). The total amount of water on Mercury is just about 1 ppb of Mercury's overall mass. This compares to 0.023 % of surface and close-to-surface water on Earth (plus the same amount of water bound to minerals in the transition zone). The water contents on Ganymede exceed those on Earth by a factor of 40. On Venus, with surface temperatures around 730 K, only gaseous water has been detected. The overall amount of water in Venus' atmosphere is just 0.002%; water is thus a minor component only, even when taking into consideration Venus' high atmospheric surface pressure of 9.2 MPa, approximately hundredfold of Earth's surface pressure.

Water ice is further present on the surface of some asteroids and is likely responsible for the comet-like appearance of several of these asteroids. Mineral hydrates found in meteorites that are fragments of asteroids provide additional evidence for the preservation of water reservoirs on this side of the snow-line. The dwarf planet Ceres, the largest object in the asteroid belt, and differentiated into a rocky core and an icy mantle, has been shown to emit water plumes into space. The outflow of water is in the order of magnitude of 1 kmol s^{-1} (Küppers 2014); 1 mol of water corresponds to 18 g. Water in the form of small ice particles distributed on the surface has been detected on the asteroid 24 Themis; 24 Themis orbits the Sun at a distance of 3.2 AU, and thus beyond the snow-line. Finally, as noted previously, a considerable amount of the material constituting comets is water ice. The largest reservoirs of water in the Solar System are presumably locked up in the giant planets (Dishoeck 2014).

What about water on exoplanets? Water has been detected in the atmospheres of exoplanets; an example is HR 8799-c, a super-Jupiter in the constellation Pegasus (Table 4.2 in section 4.2). The surfaces of the super-Earths Kepler 62-e and Kepler 62-f (section 4.1 and Table 4.1) are believed to be covered by deep oceans of liquid water. As far as giant planets are concerned, model calculations have shown that the amount of water correlates with the C/O ratio (as well as with temperature and pressure): For C/O rations close to unity, most of the oxygen is locked up in CO, while for C/O ratios ≈

0.5 the major part of oxygen is retrieved in water. However, the detection of water vapor in the outflow of the aging carbon-rich Mira variable CW Leonis, with a C/O ratio > 1 (Decin 2010), challenges this view, and we may thus suspect that water is more omnipresent than hitherto assumed.

Additional evidence for the presence of water in extrasolar planetary systems comes from a spectral analysis of the atmosphere of the white dwarf GD 61. White dwarfs are the remnants of developed type A and F stars (sidebar 2.2), and thus of stars somewhat more massive and more short-lived than the Sun. The circumstellar disk surrounding GD 61 contains oxygen in excess of what could be provided by oxidic minerals (such as silicates and "MgO"); this surplus oxygen is believed to have been delivered by water from a former exoplanet orbiting the genuine star, an exoplanet that now forms a planetary debris disk after it became disrupted when the star developed towards a white dwarf (Farihi 2013).

Water is also ubiquitous in interstellar clouds, where it mainly forms on grain surfaces (Fig. 3.2 in section 3.1), and thus becomes fed into protoplanetary disks when the clouds collapse. An example is the T-Tauri star TW Hydrae, a pre-main sequence star of 0.6 $m_{\odot}$. The star is surrounded by a protoplanetary disk, stretching from 115 to 196 AU, and containing particulate as well as gaseous water. The total water reservoir amounts to several thousand earthly oceans (Hogerheijde 2011).

By far the most common stars in our Universe are M dwarfs (Table 2.2. and sidebar 2.2). M dwarfs constitute about three-quarters of all stars. An example of an M dwarf is our nearest neighbor Proxima Centauri, an M5 star ($m = 0.123\ m_{\odot}$, surface temperature $\approx$ 3050 K) at a distance of 4.24 ly. An estimated 6% of these M dwarfs accommodate Earth-sized habitable planets. According to earlier models, these planets should be dry – as a consequence of insufficient gravitational interaction in the comparatively small (low mass) planetary disk, preventing ice grains from beyond the snow-line from becoming relocated to planets in the habitable zone. However, the radioactive decay of the comparatively short-lived aluminum isotope ^{26}Al (cf. Eq. (4.4) in section 4.2) in planetesimals beyond the snow-line may have provided sufficient heat to confine water in the form of mineral hydrates. These hydrates could then have been delivered to the planets of the M dwarf by meteoritic fragments of these planetesimals, just as part of Earth's water was supplied by meteorites.

7.3. CHEMISTRY ON TITAN

Along with Earth and, to some extent, also Mars and perhaps Venus, Saturn's moon Titan is of particular interest when it comes to the capability to initiate, develop and sustain life. Titan may therefore be considered to model a broad population of exoplanets.

Titan is the second-to-largest and the most massive moon in the Solar System, almost twice as massive as our Moon, and the only moon with an appreciable atmosphere, the surface pressure of which exceeds that of the Earth by almost 50%, while the density is 4.5 times larger than Earth's. The main atmospheric constituents are N_2 (98.4% in the stratosphere, 95% in the troposphere) and CH_4 (1.4 and 4.9%, respectively). In addition, there are traces of H_2 (0.1 to 0.2%), hydrocarbons such as acetylene C_2H_2, propyne $HC{\equiv}C-CH_3$, ethane C_2H_6, propane C_3H_8 and benzene C_6H_6, acetonitrile H_3C-CN, cyanoacetylene $HC{\equiv}C-CN$, hydrogencyanide HCN, the carbon oxides CO and CO_2, ^{40}Ar (produced, as on Earth, by β^+ decay of ^{40}K), He and cyanogen $(CN)_2$. Finally, traces of H_2O have been detected (Figure. 7.2). The haze in the atmosphere, generated from condensed oligo-aromatic hydrocarbons, masks Titan's surface in the visible, and also protects molecules in the lower atmosphere and in close-to-surface areas from destructive radiation. In analogy to water on Earth, methane (and ethane) on Titan form clouds, associated with heavy rains and surface lakes.

The formation of more complex molecules starts at high atmospheric levels exposed to short wave-length UV (< 100 nm): photolysis of N_2 and CH_4 triggers the formation of acetylene C_2H_2 and hydrogen cyanide HCN. These molecules are transported to lower atmospheric levels, where there are less energetic UV generates radicals such as C_2H, CH_3, and CN that further react to more complex molecules. End-products are refractory hetero-polymeric essentially C- and N-based macromolecules resembling tolins. These tolins are responsible for the orange-colored atmospheric haze. All in all, the chemistry taking place in Titan's atmosphere is strongly influenced by the photon flux from the Sun, and the electron flux stemming from the Saturnian magnetosphere (Raulin 2012).

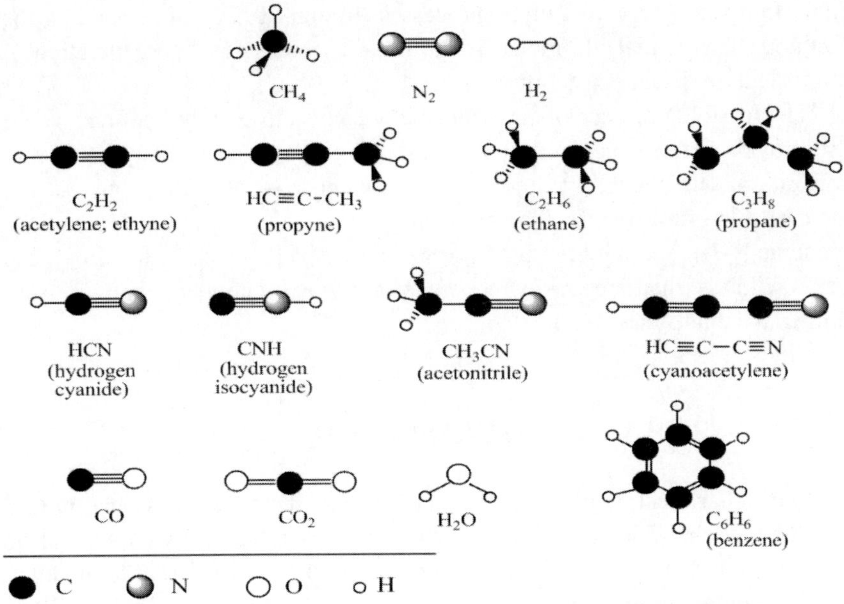

Figure 7.2. Structural and valence bond formulae (lone electron pairs omitted) of molecules identified in Titan's atmosphere, essentially based on explorations by the Cassini-Huygens probe. The chemical processes underlying the formation of the more complex representatives of these molecules may have relevance for the prebiotic formation of comparable molecules on the early Earth.

Titan's mean density is 1.88 g cm^{-3}, and hence about one-third that of Earth, indicative of a setup from a mix of low and medium density materials such as water and silicate. Titan's surface is covered with extended methane-ethane lakes, and an approximately 75 km thick ice crust composed of water and hydrocarbons, while the highlands are mainly water ice. There are widespread subsurface oceans consisting of water with some ammonia and a high salt content (potassium and sodium sulfate), comparable to that of the Dead Sea (Mitri 2014).

These oceans extend to a depth of ca. 300 km. Supposedly, there are occasional "hot spots" in the crust, allowing, along with the methane lakes, for the escape of gaseous methane, thus constantly replenishing Titan's atmosphere with CH_4, a gas that − once in the atmosphere − is comparatively rapidly disassembled by UV radiation.

A reactive water-ammonia mix, when brought to the surface, would allow for the transformation of organics present on or close to surface areas, including the incorporation of oxygen atoms into the organic molecules on

Titan, for example via hydrolytic processes (Raulin 2012), thus extending the inventory of potential life molecules to building blocks for the generation of amino acids and sugars (section 6.1).

The feasible presence of cryovolcanic hotspots close to the seafloor would also provide deep sea vents comparable, to some extent, to the black and white smokers on our planet, and providing temperatures – at least locally – that could enable primitive life. The high salt contents, the alkalinity (affected by ammonia NH_3), and the high pressure would not be an obstacle for life; these are conditions that are easily coped with by extremophiles known from terrestrial habitats (section 5.4).

7.4. PANSPERMIA

Interstellar dust grains contain and protect the molecules that render possible life (ch. 3), including amino acids, even though primitive life forms or germs thereof cannot be expected to survive in dust particles. These modules of life required for the generation of complex molecules that *feature* life, but do not yet represent life *per se*, had been supplied to – and became further processed on – planetesimals in the protoplanetary system, and consequently became delivered to the planets that formed by aggregation of the planetesimals. Those of the planetesimals that were left over, the asteroids and the dwarf planet Ceres in the asteroid belt of our system, still witness this early scenario by delivering splinters to Earth: meteorites, deriving from the meteoroids formed by (collisional) fragmentation of asteroids. Life molecules thus have been carried to the planets also at later stages of the planets' evolution; but to emphasize it once again, "life" is not identical to molecules which enable life.

In Chapter 5 (and in section 5.2 in particular) we have pointed to the fact that the emergence of life – at least of life in the form with which we are acquainted – has likely been a singular event in our galaxy, given that the multitude of processes that have to proceed in order to generate Life is complex to an extent that probability severely diminishes the feasibility for life to emanate at all. This view raises (at least) two question: (i) did life originate on Earth, and might it have spread to other planets and moons, including extrasolar systems; and (ii) did life originate somewhere else in our galaxy and become conveyed to Earth? There is (at least presently) no way to unravel this conundrum. And consequently, the question of whether and how life might have become disseminated across our Solar System, across the neighborhood of the Solar System, or even the Milky Way Galaxy and the Local Group[34] has been subject to faith and belief, esotericism, and science fiction, but to scientifically based consideration as well. Pseudo-religious instrumentalization in particular has contributed to discrediting the scientific approach to panspermia.

The transmission of primitive life forms – for example, microbes, or pre-life forms such as represented by the RNA world (*cf.* Figure 5.1) – to habitable celestial bodies is commonly referred to as panspermia, a term that goes back to the Greek philosopher Anaxagoras (500-428 BC) and translates as "seeds everywhere". According to Anaxagoras, all things originally existed in infinitesimally small fragments of themselves (the seeds, or spermata) that had to be eliminated from the *complex mass* before they could receive a *definite character*. From a contemporary point of view we may interpret "complex mass" as *celestial bodies*, and "definite character" as *(proto-)form of life*. At the beginning of the eighteenth century, this idea became revived by the French Government official Benoit de Maillet, who wrote that "The entire space is full of [delicate and minute] seeds of everything which can live in the Universe" (Wainwright 2010).

The notion of panspermia became revived, in part on a scientifically justified basis, in the nineteenth century, essentially as a consequence of new insight into, and awareness of, the origin and composition of meteorites, or as Lord Kelvin (known for his contributions to thermodynamics and electrodynamics) put it in 1871, "seed-bearing meteoritic stones moving about through space". Other early protagonists of panspermia include Berzelius and Helmholtz. A first detailed account on panspermia became issued by the Swedish chemist and physicist Svante Arrhenius, known for his contributions to kinetics and to the dissociation of ionic compounds (electrolytes) in water, and winner of the 1903 Nobel Prize in chemistry. At the beginning of the 20th century, Arrhenius postulated that spores (of bacteria, protozoa, algae, fungi and plants) can be transported off Earth into space by the radiation pressure exerted by the solar light, and he also proposed that bacterial spores, propelled through space by the light pressure of stars, could have been the seeds of life on our planet. Arrhenius did not simply theorize, but backed up his assumptions by calculations. At that time, the solar wind had not yet been discovered. The solar wind is a highly energetic stream of mostly protons and

---

[34] Along with the Milky Way, the Local Group encompasses a few dozen additional galaxies and dwarf galaxies. The Andromeda nebula, and the Large and Small Magellanic Clouds are part of the Local Group.

electrons emitted by the Sun, and nowadays considered to function as an alternative potential transport medium. More recent proponents of panspermia include Francis Crick, Fred Hoyle, Leslie Orgel, Steve Hawkins, and Chandra Wickramasinghe. Wickramasinghe in particular demonstrated that carbonaceous materials detected in interstellar dust grains can plausibly be explained as arising from the degradation of biologically derived molecules (Rauf 2010).

Life on Earth developed to high standards, but this fact just reflects that the conditions for the evolution, adaption and perhaps perfection of life have been particularly favorable on our home planet. It does not necessarily imply that life started on Earth. Our neighboring planet Mars and, as outlined in section 7.3, Saturn's moon Titan, may well have been the birthplace of life in our planetary system. Martian meteorites collected on Earth clearly indicate that through-space transport − local panspermia, so to say − might have provided Earth with early pre-life forms of life that, once accommodated here, came upon more favorable conditions than on Mars or Titan for further development. Supra-pure magnetite Fe_3O_4, embedded in carbonates $(Fe,Mg)CO_3$, detected in the Martian meteorite ALH84001 had been interpreted in terms of the presence of magnetotactic bacteria[35] on Mars some 4.1 billion years ago (Thomas-Keprta 2009), although this interpretation is still being disputed.

Hoyle and Wickramasinghe also proposed that comets could transport bacteria across the *galaxy*, shielding bacterial life from destructive radiation by embedment in the comet's icy interior. In our system, the main source of comets is the outer Oort cloud at a distance of about 100.000 AU (Figure 2.1).

Disturbances caused, for example, by our next neighbors, the fixed stars α Centauri and Proxima Centauri, can deflect comets into orbits of high ellipticity, or even onto a hyperbolic track, allowing the comet to travel to solar systems in our neighborhood and further into our galaxy. In a comparable manner, rogue comets from distant solar systems can eventually become subjected to the gravitational forces exerted by our Sun and thus be trapped within our system, providing a potential trail for the delivery of sperms of life.

---

[35] The tagging ALH84001 designates the first (001) meteorite found in the Alan Hills (ALH) field in Antarctica in 1984 (84). Magnetotactic bacteria are bacteria that synthesize microcrystals of magnetite for orientation in Earth's magnetic field. The early Mars had available a magnetic field as well. Suprapure magnetite is sometimes considered a biomarker.

An alternative route for panspermia has been mentioned earlier in the context of Steppenwolf planets (sections 2.2, 3.1 and 7.2), i.e., planets that escape their sun, again on a hyperbolic trail, in a comparatively early state of the formation of a planetary system.

Massive disturbances, for example, in a binary star system, or by high-mass "hot Jupiters", can provoke this escape. Such a rogue planet may be equipped with an ocean underneath a protective crust, similar to the situation encountered with Titan or Enceladus, where the development of primitive life is nourished by deep-sea hydrothermal vents, fed by the magma of the planet. Certainly, these are adventurous or even quixotic scenarios, at least at the present stage of comprehension of our Universe.

SUMMARY

Habitability – the development and sustenance of life – is a matter of moderate temperatures (on a planet or moon), allowing for the presence of, at least locally, liquid water, and the persistence of organic matter. But even for planets orbiting their sun in the habitable zone, habitability can be counteracted by the locking of the planet's rotational motion to its orbital motion, and/or the loss of obliquity. Moons profiting from tidal heating by their planet and/or from internal volcanism can eventually harbor sub-surface oceans. Likewise, rogue planets, expelled from their home system shortly after planet formation, may resort to subsurface oceans.

Water is omnipresent in the Universe, and the deposition and thus retention of water ice on dust grains in interstellar clouds renders available this water when planets form in the course of the collapse of these clouds. Planets, however, forming within the snowline are not as easily supplied with genuine water. In this case, secondary processes, such as the capture of comets and meteorites, as well as the retention of water in the form of mineral hydrates and of hydroxides, can provide *those* planets with water that orbit their sun inside the snowline.

Saturn's moon Titan may be considered a model for several features intrinsic to exoplanets: Titan is equipped with a comparatively dense atmosphere (mainly N_2 and CH_4) and methane-ethane lakes, with a crust of (water) ice, and a salty alkaline subsurface ocean. Titan's atmosphere is venue for the formation of various carbon- and nitrogen-based organics; the subsurface ocean might be a birthplace for more complex molecules, nourishing speculations that, in the vicinity of hotspots in the subsurface sea, primitive life or pre-life forms have developed.

Life is unique to the extent that it probably *developed* only once in the Universe, minimizing the chance that life exists elsewhere. This restriction does not exclude the possibility that life is also *present* elsewhere in the Universe. Independently of whether life developed on Earth or somewhere else in the Solar System or beyond, primitive life forms or spores thereof might have been transported through space and relocated on habitable bodies, a notion known as panspermia. The perception of "panspermia" goes back to the Greek philosopher Anaxagoras, who lived in the 5th century BC; the idea became revived from the 19th century AC. "Modern" protagonists of panspermia include Arrhenius and, more recently, Wickramasinghe, the main proponents of scientifically-based suggestions for the through-space transportation of germs of life.

REFERENCES

Anand, M. (2014): Analyzing Moon rocks. *Science* 344:365-366.

Barlow, M.J., Njuyen-Q-Rieu, Truong-Bach, et al. (1996): The rich far-infrared water vapour spectrum of W Hya. *Astron. Astrophys.* 315:L241-244.

Chabot, N.L., C.M. Ernst, B.W. Denevi, et al. (2014): Images of surface volatiles in Mercury's polar craters acquired by the MESSENGER spacecraft. *Geology* doi 10.1130/G35916.1.

Christner, B.C., J.C. Priscu, A.A. Achberger, et al. (2014): A microbial ecosystem beneath the West Antarctic ice sheet. *Nature* 512:310-313.

Decin, L., M. Agúndez, M.J. Barlow, et al. (2010): Warm water vapour in the sooty outflow from a luminous carbon star. *Nature* 467:64-67.

van Dishoek, E.F., E.A. Bergin, D.C. Lis, and J.I. Lunine (2014): Water: from clouds to planets. *Astrophys. Galaxies* 01/2014.

Gollihar, J., M. Levy, and A.D. Ellington (2014): Many paths to the origin of life. *Nature* 343:259-260.

Farihi, J., B.T. Gänsicke, and D. Koester (2013): Evidence for water in the rocky debris of a disrupted extrasolar minor planet. *Science* 342:218-220.

Finkelstein, S.L., C. Papovich, M. Dickinson, et al. (2013): A galaxy rapidly forming stars 700 million years after the Big Bang at redshift 7.51. *Nature* 502:524-527.

Fischer, E., G.M. Martínez, H.M. Elliot, and N.O. Rennó (2014): Experimental evidence for the formation of liquid saline water on Mars. *Geophys. Res. Lett.* 41:4456-4462.

Heller, R., R. Barnes, and J. Leconte (2011): Habitability of extrasolar planets and tidal spin evolution. *Orig. Life Evol. Biosph.* 41:539-543.

Heller, R., and J. Armstrong (2013): Superhabitable worlds. *Astrobiology* 14:50-66.

Hogerheijde, M.R., E.A. Bergin, C. Brinch, et al. (2011): Detection of the water reservoir in a forming planetary system. *Science* 334: 338-340.

Keller, C.U. (2012): In search of biosignatures. *Nature* 483:38-39.

Küppers, M., L. P'Rourke, D. Bockelée-Morvan, et al. (2014): Localized sources of water vapour on the dwarf planet (1) Ceres. *Nature* 505:525-527.

Loeb, A. (2014): The habitable epoch of the early Universe. *Int. J. Astrobiol.* 13: 337-339.

Lucey, P.G. (2013): A wet and volatile Mercury. *Science* 339:282-283.

Mitri, G., R. Meriggiola, A. Hayes, et al. (2014): Shape, topography, gravity anomalies and tidal deformation of Titan. *Icarus* 236:169-177.

Rauf, K., and C. Wickramasinghe (2010): Evidence for biodegradation products in the interstellar medium. *Int. J. Astrobiol.* 9:29-34.

Raulin, F., C. Brassé, O. Poch, and P. Coll (2012): Prebiotic-like chemistry on Titan. *Chem. Soc. Rev.* 41:5380-5393.

Thomas-Keprta, K.L., S.J. Clemett, D.S. McKay, et al. (2009) Origins of the magnetic nanocrystals in Martian meteortite ALH84001. *Geochim. Cosmochim. Acta* 73:6631-6677.

Wainwright, M., and F.A. Alshammari (2010): The forgotten history of panspermia and theories of life from space. *J. Cosmol.* 7:1771-1776.

Chapter 8

EPILOGUE

According to a presently widely accepted theory, our Universe came into existence 13.8 billion years ago, by an event termed "Big Bang" (a concept originally coined by the astronomer and science fiction writer Fred Hoyle to discredit this theory). According to the Big Bang theory, the Universe expanded within fragments of a second (the inflation period) from a singularity (a single point of almost zero extension) towards a spatial extent not much different from today's spread. The Universe is still expanding, though at a substantially reduced rate. This origin and development of the Universe was originally proposed by Lemaître in 1927, but can actually be traced back to the year 1225: The Big Bang theory, or at least what comes close to the modern understanding of the origin and development of the Universe, has already been delineated about 800 years ago: In his œvre *De Luce*, the cosmologist Robert Grosseteste proposed that "an initial explosion of sort of a light expanded the Universe into an enormous sphere, thinning matter as it goes" (McLeish 2014).

Within the epoch of 10-17 million years after the Big Bang, overall temperatures in the infant Universe were moderate to the extent where liquid water and Life as we know it might have existed (section 7.1) – though this stretch of time should not have sufficed for the *development* of life. Realistic conditions for life to evolve and to develop became available with the formation of the first stars and their planetary systems. So far the oldest of these objects date back to 700 million years after the Big Bang, hence a bygone era. Our Solar System came into existence 4.57 billion years ago (9.2 billion years after the Big Bang), and it took approximately another 0.7 billion years for life to develop on Earth – or to be inseminated via panspermia (section 7.4).

The emergence of life and its proliferation in space has been addressed in science fiction and likewise in science. An interesting treatise in this respect became introduced almost nine decades ago by the Swedish astronomer Knut Lundmark in his book Världsrymdens Liv (Lundmark 1926), literally translated into "Life in Space", in which the author addresses the issue of the principal alternatives as to how life might have developed and spread out, stating that life has existed since time immemorial, or evolved from inanimate matter either in many places, or just once, followed by migration to other worlds.

In the 1970s, James Lovelock and Lynn Margulis came forward with an interesting hypothesis according to which Earth as a whole is a living organism, in the sense that habitability is a direct consequence of Earth functioning as a self-regulatory system in which the abiotic environment is in intimate interaction with the biosphere, thus sustaining the overall conditions for life. This approach has become known as the "Gaia hypothesis"; the goddess Gaia is, in the Greek mythology, the personification of Earth. The Gaia hypothesis, although still controversially discussed, becomes increasingly attractive in the context of environmental issues that influence and thus modify the conditions for life (Lovelock 2014).

As for many other planetary systems that emerged earlier, or at about the same time as our system with, however, sufficiently more massive suns, there will be no happy ending: When our Sun has developed off the main branch into a red giant within the oncoming one billion years, temperatures on Earth will rise to the extent that Life will become extinct – long before Earth is finally devoured. With the aging Sun, conditions for life on Earth will become increasingly hostile, and life will likely return to unicellular forms. The microbial life could persist for approximately another 2.8 Gy from the present, before it becomes extinct for good (O'Malley-James 2014).

Earth-like exoplanets in the habitable zone of their Sun-like (G-type) stars are, statistically seen, likely to be in an early or in a late phase of their development; it is therefore more likely to encounter unicellular rather than multicellular life on these planets. In contrast, the conditions for multicellular life on Earth-like planets orbiting stars that are less massive (K- and M-type stars) than our G2 sun are sufficiently more profitable in the sense that the time-span for multicellular organisms to develop and to subsist is much more extended.

On these exoplanets we might expect to encounter multicellular life that boasts a higher level than on Earth – provided, of course, that life exists at all on other planets. I have pointed to this caveat in section 5.2.

On exoplanets orbiting aging suns we cannot expect to encounter life. Other exoplanets are not (yet) in a predisposition enabling the development of life, because they have just evolved from collapsing interstellar clouds, or because they are too close to their sun to render possible the constitution of more complex organic matter, and thus to initiate the germination of life.

Exoplanets may reside in a zone where biological activity is already extinct ("burnt up"), or where the development of biological activity is out of range. Nonetheless, millions of exoplanets in our Milky Way Galaxy (and, of course, in other galaxies as well) can be expected to have encountered situations comparable to Earth, and hence to experience habitable conditions similar to those persistent to our home planet. But how do we get access to information that provides "biosignatures" – evidence that relates to living organisms?

I have briefly pointed to the potential for detecting biosignatures in section 7.1. To date, the availability of the respective information is minute, however. Certainly, the situation will improve with the progress in instrumentation employed in the search and characterization of exoplanets. Future projects and missions dedicated to the search of exoplanets include:

- GPI (Gemini Planet Imager) in Cerro Pachon, Chile, operated by a North American consortium since mid-2014
- SPHERE (Spectro-Polarimetric High-contrast Exoplanet REsearch), operated by a European consortium at the Paranel Observatory in the Chilean Andes as of mid-2014
- CHEOPS (CHaracterizing ExOPlanet Satellite), exoplanet mission by the ESA, projected start in 2017
- TESS (Transiting Exoplanet Survey Satellite), exoplanet mission by the NASA, projected for 2017
- James Webb Space Telescope (successor of the Hubble and the Spitzer Space Telescopes), focusing on IR observations directed towards the formation and evolution of galaxies, stars and planetary systems; cooperated by NASA, ESA and CSA
- PLATO (PLAnetary Transits and Oscillation of stars), exoplanet mission by the ESA, planned for 2024.

REFERENCES

McLeish, T.C.B., R.G. Bower, B.K. Tanner et al. (2014): History: A medieval multiverse. *Nature* 507:161-163.

Lovelock, J. (2014): *A rough ride to the future.* Penguin Books, UK.

Lundmark, K. (1926): *Världrymdens Liv. Lindblads Förlag,* Uppsala.

O'Malley-James, J.T., J.S. Greaves, J.A. Raven, and C.S. Cockell (2012): Swansong biospheres: Refuges for life and novel microbial biospheres on terrestrial planets near the end of their habitable lifetimes. *Int. J. Astrobiol.* 12:99-112.

INDEX